Dream On!:
Supporting and Graduating African American Girls and Women in STEM

Ezella McPherson, Ph.D.

Foreword by Alexa Canady-Davis, M.D.

Written by: Ezella McPherson © 2021

All Rights Reserved. No part of this publication may be reproduced, stored in a retrieval system, or transmitted in any form or by means of electronic, mechanical, photocopying, recording or otherwise by any means without the written permission of the author.

Disclaimer: The names of the actual people have been changed to pseudonyms out of respect for their professional identities. The neighborhoods, schools and colleges, and cities have been changed to pseudonyms as well.

Cover illustrations by Sabreta Kennedy

Printed in the United States of America

ISBN: 978-1-7372731-2-7

Dr. McPherson Coaching, LLC
Detroit, Michigan
www.drmcphersoncoaching.com
emcpher2@gmail.com

Library of Congress Control Number: 2021918411

1. African American women **2.** African American girls **3.** Science and Math **4.** STEM Education **5.** Hidden Curriculum **6.** Teaching and Learning **7.** Social Capital **8.** Graduation **9.** Family

Advance Praise for

Dream On!: Supporting and Graduating African American Girls and Women in STEM

"*In this thoughtful and timely work, Dr. Ezella McPherson provides fresh insights into the types of experiences that can help young women of color complete STEM majors or else steer them to other pathways.*"

Dr. Freeman A. Hrabowski III
President, University of Maryland Baltimore County
Author of *Overcoming the Odds: Raising Academically Successful African American Young Women* (with Kenneth I. Maton, Monica L. Greene, and Geoffrey L. Greif).

"*It is curious that a majority of Americans support the American Dream and equal opportunity for all but look the other way when hidden curriculums create blatant racial inequality in STEM education. In the book, Dr. McPherson uses the voices of young African American women to provide insights into the processes at work and to suggest recommendations for teachers and practitioners.*"

Dr. Sandra L. Hanson
Professor Emerita, Department of Sociology, Catholic University of America
Author of *Swimming against the Tide: African American Girls and Science Education* and *The American Dream in the 21st Century* (with John K. White).

Table of Contents

Dream On!: Supporting and Graduating African American Girls and Women in STEM

Foreword by Dr. Alexa Canady-Davis i

Acknowledgements iii

Dedication v

Preface vii

Introduction 1

Chapter 1: African American Girls' Experiences in Science and Math Classrooms in K-12 settings 15

Chapter 2: African American Girls' Student Engagement in Science and Math in K-12 Schools 23

Chapter 3: Teaching and Learning Science and Math in K-12 School Settings 37

Chapter 4: The Hidden Curriculum Within the STEM Culture in Higher Education 61

Chapter 5: Invisible in High School and College: African American Girls and Women's Support Networks 83

Chapter 6: Lost Talent in the Leaky Pipeline: African American Women Cooled-Out From STEM Majors 105

Chapter 7: Dream On!: African American Women's Thoughts on College Completion 123

Chapter 8: Recommendations for STEM Student Success for African American Girls and Women..........................131

Chapter 9: Conclusion and A Call to Action for African American Girls and Women to Keep on Dreaming in STEM and Medicine..157

Epilogue: Case Profiles and Where are They Now........159

Appendix: Table 3: Overview of the Participants..........197

Biography..201

Foreword

Barriers exist for all people daring to seek careers in STEM. For women and minority women, tremendous personal strength is necessary to dare to dream the dream, to persist despite the absence of external encouragement and often active discouragement. The student's ability to believe in that framework that their dream is possible is already a major accomplishment. Many minority women are further burdened by marginal schools. In a field that is built brick on brick like mathematics and the hard sciences, even one bad teacher can weaken the wall. Once I was tutoring an international baccalaureate student in calculus; it became clear that she fully understood the calculus concepts, but that her advanced algebra skills made it impossible to calculate the problems. Only if you think the dream is achievable, then you are willing to make all the sacrifices that are necessary for successful STEM careers.

For many, the family provides that reinforcement in addition to those favorite teachers that saw your potential. For minority students, although parents would love to celebrate their child's success, their enthusiasm is tempered by the recognition of the almost impossible barriers and the desire to help their child avoid the devastation that follows failed dreams knowing the odds are against them. We cannot overlook the importance of a peer group in student success as well. For example, in general surgery programs that had a cohort of women who were successful derived support from a sense of community.

However, many of my classmates had stories of teachers who told them they were reaching to high; that college was not a reasonable goal for them. My husband, a graduate of Pepperdine University, was told in high school that he was not college material. He was able to turn that pronouncement into fuel for succeeding. Additionally, many students have reached out to me over the years expressing their desire to be neurosurgeons. I enjoy interacting and encouraging them knowing full well that most will not be neurosurgeons. But if you want to be something, whatever it is initially you are highly likely to become something good. I wanted to be a mathematician, but I didn't fit in that world and so I became a neurosurgeon, a pretty good outcome.

Listen to the stories of the women in the book to understand how they experienced their education and use these lessons to help you engage and successfully mentor the students that come into your life.

Dr. Alexa Irene Canady-Davis
Retired, neurosurgeon

Acknowledgements

First and foremost, I'd like to thank God for allowing for me to complete my middle school dream of publishing a book. I am grateful for leaning on him for guidance throughout the publishing journey. I also want to thank my family, friends, and mentors for their support. Thank you for cheering me on throughout this process. I would never have made it this far without the support of the sensational 16, the ladies who participated in this research study. Although their names are pseudonyms in the book, I am grateful to them for sharing their stories of trials and tragedies, as well as triumphs in science, technology, engineering, and math (STEM) majors. I also want to thank the first African American woman neurosurgeon, Dr. Alexa Irene Canady-Davis for supporting this book as well by reviewing the book and writing the foreword. I'd also like to thank my endorsers, Dr. Freeman Hrabowski III and Dr. Sandra Hanson for reviewing the book as well. Finally, I'd like to thank my undergraduate friends who inspired me to think of and write a book on the lived experiences of African American women in STEM majors. It is through you that I learned first-hand some of the challenges that occur while pursuing STEM majors and pre-medicine degrees.

Dedication

This book is dedicated to all of the African American women who have ever dreamed of pursuing a STEM degree or a pre-medicine course of study. The book is also dedicated to African American women who have earned STEM degrees and medical degrees. You inspire us all to live out our dreams despite obstacles.

Preface

Growing up as an African American girl in Norman, Oklahoma, I was a dreamer. In elementary school, my career aspiration was to become a nurse. At that time, my father worked as a nurse's assistant at the local hospital, Norman Regional Hospital. My father worked at the hospital at about the same time as I had my first African American teacher, Mr. Thatcher. He was the physical education teacher at Lincoln Elementary School, and we often referred to him as Coach Thatcher.

In sixth grade, I dreamed of attending the University of Michigan-Ann Arbor. However, one of the reasons I eventually gave up on my career aspiration of being a nurse was that I lacked role models who were African American woman in the nursing field. I also despised the thought of seeing blood and needles on a daily basis, so a career in nursing became unlikely. By that time, my father had left his position at Norman Regional Hospital as the nursing assistant. In addition, in sixth grade, I took a math class from Mrs. Thatcher (Coach Thatcher's wife). She was a biracial woman and my only African American woman math teacher throughout my primary and secondary schooling. In retrospect, I began liking math in part because of her warm classroom environment and caring instruction. Similarly, my eighth-grade science instructor, Mr. Walker, became my third and final African American instructor throughout my middle school and high school years. I never really liked science until Mr. Walker came along and made science fun, engaging, and interactive. He became a role model for minority students as well. My subsequent high school science instructors were older White men.

By high school, I had fallen in love with math even though I was exposed to the chilly culture of science. The culture of science is a term used to describe the environment associated with the teaching and learning of science. The culture of science is often taught through a Western mode of thought that privileges the contributions of White men (Lee & Luykx, 2006). The culture of science also has a competitive culture that thrives on competition within science, technology, engineering, and math (STEM) fields. I first experienced the culture of science in my ninth grade algebra class with Norman High School's football coach, Mr. Franklin, a White man. He handed back exams based on grades. Students who earned an A got their exams back first, and those who received a D or an F got theirs last. He made learning math a competition. I vividly remember that he announced to the class, "Only one person failed the exam." He handed back my exam, second to last. Soon thereafter, I flunked my first algebra exam, and I was devastated. My sister, who at the time was in the same class as me, made a better grade than me on the first exam. Learning algebra was hard because it became a competition to win in the class. Winning meant getting my exam handed back first versus last. Then, on the next exam I made a D and then another D. I studied harder and pulled a C on an exam. I also went to algebra study hours after school. My hard work paid off, and I began making B's and A's on exams. I was so grateful to be one to survive this competitive math culture of the survival of the fittest by making higher grades. I made a solid C my first semester of Algebra. That was the only C on my high school transcript.

In 10th grade, I took geometry with a White female instructor, Ms. White, who was also the women's basketball coach. Although the climate was different in the math class, I could not understand the materials the way she taught them. So, I ended up with a solid B in her class. My 11th grade math class was Algebra II, and in 12th grade, I took trigonometry/advanced algebra to prepare for college entrance exams. By that time, I knew that I wanted to major in business. I even took a personal finance class in high school. I also observed that I was one of two African American women in the trigonometry/advanced algebra class taught by a White female instructor.

Since I was college bound, I took chemistry during my senior year. I was the only African American woman in my class. Chemistry was difficult, but I used my resources, mainly the instructor and student teacher, and successfully passed the class with a solid B. I also experienced firsthand the feeling of being isolated because of the culture of science in the classroom. I rarely participated in class unless the teacher called on me. There were problems with the math and science cultures, including the lack of racial diversity among my science and math teachers. In addition, few students of color took advanced placement or honors classes or even advanced math (e.g., advanced algebra, trigonometry, precalculus) or chemistry to prepare for college. This meant that few of the students of color in my school, especially African American women were prepared for the rigors of college math or science.

My middle school dream of attending the University of Michigan-Ann Arbor became a reality when I entered as a freshman who desired to major in business. I was a part of the Residential College, which afforded me the opportunity to take smaller classes with my peers. During my freshman year, I also had friends who dreamed of becoming doctors, businesswomen, engineers, or lawyers and sought other types of professional careers. My career aspiration changed to become a businesswoman. However, I learned about the rigors of the curriculum in the University of Michigan Stephen M. Ross School of Business through upperclassmen in the University of Michigan National Association for Black Accountants (NABA), which was an undergraduate organization. As a result, my thoughts about business changed, and the business major no longer appealed to me.

In college, I took Statistics 100 from an international woman instructor. She had a heavy accent and did not teach statistics in a way that I could fully learn it. I pulled a B- in that class. During my junior year of college, I enrolled in Statistics 350 with a White female instructor. She was knowledgeable about the course materials and presented them in a way that made it easy for students to learn. However, family issues affected my performance in that class, and I made a C- in that class. That was the only C grade on my undergraduate transcript.

Since math was a challenge for me in college, I could tell that something had happened, and my interest in math had waned. My high-achieving African American female friends at the University of Michigan-Ann Arbor had lost interest in math and science as well. I also witnessed first-hand the challenges that my friends who wanted to become doctors encountered in chemistry and mathematics classes at the University of Michigan. The aftermath of the math and science curricula at the University of Michigan resulted in the majority of my undergraduate friends also placing their dreams on hold. However, one friend became an engineer (Akiya Jones), two friends became doctors (Dr. Dominique Adams Hill and Dr. Chinwe Nwosu), one friend is now a nurse (Narene Hyman Beachem), one friend (Ashley Thomas, now deceased) earned a Bachelor of Science degree in Statistics, and two friends (Antoinette Price and Shannon Wilson) obtained Master of Public Health (MPH) degrees. Some of my

African American women friends transitioned into health science majors. My life story suggests that there is a small pool of African American women left in STEM fields and pre-medicine in college.

Nearly a decade after college graduation, the two college African American women friends who earned MPH degrees are now medical doctors (Dr. Shannon Wilson Bradley and Dr. Antoinette Price). Dr. Wilson Bradley graduated with a Doctor of Medicine degree from the University of Illinois at Chicago College of Medicine in 2016. I also had an African American woman mentee, Dr. Ngozi Emuchay who graduated from the same medical school as my friend in 2016. Dr. Antoinette Price earned a Doctor of Medicine degree from the same medical school in 2019 as well. A third African American woman college friend, Dr. Diamond Moore Shelby, earned her Doctor of Medicine degree from Wayne State University's School of Medicine in 2019.

Similarly, in my career as an Academic Advisor in Wayne State University's College of Engineering, I observed few African American women enter college and earn Bachelor of Science degrees in Engineering. In this position, I certified only one African American woman for a Bachelor of Science degree in Electrical Engineering from 2013 to 2014. Additionally, as the Director of the Titan Success Center at Indiana University South Bend, I saw even fewer African American women pursue STEM degrees or pre-medicine from 2015 to 2017.

In my personal and professional experiences, I observed an underrepresentation of African American women in STEM fields and pre-medicine. Current scholars continue to note a lack of African American women in STEM fields (Ireland et al., 2018; Leath & Chavous, 2018; National Science Foundation, 2019). The rationale for this book is thus to better understand why so few African American women are earning bachelor's degrees in STEM fields. This work examines the primary and secondary school successes and preparation of African American women who sought STEM degrees at a predominantly White institution (PWI) in the Midwestern part of the United States, identified herein as Town University. The text also explores the challenges these students faced while pursuing STEM majors at this PWI. In addition, the book discusses the student support networks that African American girls and women can use to navigate through STEM majors in college. Recommendations for STEM student success are provided as well.

In the Introduction, readers learn about the problems of a STEM culture dominated by White men and the underrepresentation of African American women in STEM fields. To better understand African American women's experiences in math and science in girlhood, Chapter 1 asks the question "Does early exposure to science or math influence African American girls' engagement in STEM?" This chapter presents African American women's reflections on their girlhood experiences of early exposure to science and math in primary and secondary school settings through teachers who engaged them in science and math projects. This exposure to science helps to develop the science identities of African American girls who want to later pursue science as a major in college or career.

Chapter 2 explores the influence that teachers have on African American girls' engagement in STEM. Specifically, the study participants reflect on their childhood experiences of leadership and participation in science and math classrooms and then describe their engagement in science projects and involvement in enriched curricula, more specifically, advanced placement and honors courses.

Chapter 3 examines the teaching styles that enable African American girls to learn math and science. This chapter considers the question "What are the science and math teaching methods that promote student engagement for African American girls?" African American women also reflect on the culturally responsive teaching and caring instruction that they received as girls to engage with and learn math and science in warm classroom environments despite the lack of racial diversity in teacher demographics. This chapter also presents African American women's reflections on their preparation for science and math in college.

Chapter 4 introduces the hidden curriculum within the STEM culture. This chapter addresses the question "What is the hidden curriculum within STEM majors at Town University?" Readers learn about the hidden curriculum within the STEM culture that discourages African American women from pursuing STEM majors at a PWI. The hidden curriculum in STEM majors revolves around (1) academic rigor and competitiveness, (2) course failure and information overload, (3) attendance at office hours, (4) making use of campus resources, and (5) departmental culture. This hidden curriculum creates racial and gender inequalities with regards to equal opportunities to learn in STEM.

Chapter 5 discusses the support networks that African American girls and women use to navigate through STEM fields in secondary and postsecondary settings. Specifically, the social capital framework is employed to explain how the use of collective resources (e.g., parents, professors, academic advisors, mentors, peers) provides African American girls and women with social capital in the form of information and resources that show them how to be successful in STEM, social science, and health science majors at Town University.

Chapter 6 argues that African American women leave STEM majors in college as a result of being pushed out during the five-phase cooling-out process. The five phases of the cooling-out process (Clark, 1960, 1980) are denoted as (1) alternative achievement, (2) gradual disengagement, (3) objective denial, (4) consolation, and (5) avoidance of standards. Chapter 7 investigates the aspirations of African American women to complete college. Specifically, African American women students at Town University describe their personal, professional, and educational reasons for completing STEM, social science, and health science majors.

Chapter 8 provides recommendations for helping African American girls and women to be successful in STEM. Recommendations include: (1) modifying teaching methods to engage African American girls and women in STEM fields in K–16 settings, (2) exposing the hidden curriculum in STEM, (3) providing student support networks for African American women in college, and (4) implementing warmer departmental climates to facilitate their graduation from college. Chapter 9 discusses the conclusions of the chapter and gives a call to action to encourage African American girls and women to continue dreaming and pursuing their dream career goals in STEM majors and medical degrees. Finally, the epilogue provides case profiles and describes where the participants are now in their education and careers.

References

Clark, B. R. (1960). The cooling-out function in higher education. *The American Journal of Sociology, 65*(6), 569-576.

Clark, B. (1980). The "cooling-out" function revisited. In G. Vaughn (Ed.), Questioning the community college role. *New Directions for Community College, 32,* 15-31.

Ireland, D. T., Freeman, K. E., Winston-Proctor, C. E., DeLaine, K. D., Lowe, S. M., & Woodson, K. M. (2018). (Un)hidden figures: A synthesis of research examining the intersectional experiences of Black women and girls in stem education. *Review of Research in Education, 42,* 226–254. https://doi.org/10.3102/0091732X18759072

Leath, S. & Chavous, T. (2018). Black women's experiences of campus racial climate and stigma at predominantly white institutions: insights from a comparative and within-group approach for stem and non-stem majors. *The Journal of Negro Education, 87*(2), 125-139. https://www.jstor.org/stable/10.7709/jnegroeducation.87.2.0125

Lee, O., & Luykx, A. (2006). *Science education and student diversity.* New York: Cambridge University Press.

National Science Foundation. (2019). *Minorities, women, and persons with disabilities in science and engineering: 2019.* (Report No. 19-304). https://ncses.nsf.gov/pubs/nsf19304/

Introduction

Historically, African American women have earned bachelors' degrees from both Historically Black Colleges and Universities (HBCUs) and predominantly White institutions (PWIs) (Bush et al., 2009; Evans, 2007; Jordan, 2006; McPherson, 2013; Warren, 2000). The first African American female college graduate, Lucie Stanton Day, earned a literary degree from Oberlin College, a PWI, in 1850 (McPherson, 2013). Oberlin College was the first college to admit women (Evans, 2007; McPherson, 2013). During the 19th century, the majority of Black women who received college degrees earned them from HBCUs. Almost a century after the first Black woman graduated from college, African American women were earning about twice as many college degrees as African American men in the 1940s (Blalock & Sharpe, 2012; Bush et al., 2009). As a result of World War II, African American women were afforded opportunities to earn college and graduate degrees in science, technology, engineering, and math (STEM) fields (Brown, 2011; Jordan, 2006; Warren, 2000).

African American women had limited access to PWIs before the 1970s as well. Since the 1970s, Black women have increased their rates of participation and graduation from all four-year colleges (Blalock & Sharpe, 2012; Bush et al., 2009; Gold, 2011; Robinson & Franklin, 2011; U.S. Department of Education & National Center for Education Statistics, 2019). There are several reasons for this trend. After high school, some African American women delay postsecondary education or attend community colleges prior to enrolling in four-year institutions (Dixson & Chambers, 2009; Jez, 2012). In addition, African American women attending the most selective colleges are more likely to graduate from college than their counterparts who matriculate at less selective institutions (Bowen, 2009; Walpole, 2009). The top high school students are most likely to get admitted to elite colleges and are also most likely to graduate because they excel academically.

In addition, only small numbers of Black women graduate in STEM majors. For instance, women earned just under 50% of the 666,157 science and engineering bachelor's degrees awarded in 2016. However, in this same year, African American women obtained only 33,700 (10%) of the 330,883 science and engineering bachelor's degrees awarded to women (National Science Foundation, 2019). Finally, it is important to note that HBCUs produced higher numbers of undergraduate African American women STEM degree recipients than PWIs (Perna et al., 2009). HBCUs also graduated the highest number of African American medical doctors (Gasman et al., 2017).

Culture of Science

The field of science has historically been dominated by White men (American Association of University Women, 1992; Brown, 2011; Dortch & Patel, 2017; Hill et al., 2010; Jordan, 2006; Malcom et al., 1976; Malcom & Malcom, 2011; National Science Foundation, 2019; Ong et al., 2011). This circumstance has made it difficult for many women to pursue an education or career in science, often forcing them to choose between their careers and their families (Malcom et al., 1976). The culture of science centers on teaching and learning science through a Western lens that privileges the contributions of White men to science, including their methodologies and scientific discoveries achieved through research (Committee on Barriers and Opportunities in Completing 2-Year and 4-Year STEM Degrees et al., 2016;

Harding, 2006). This institutional culture of science differs from the home cultures of some students of color (Lee & Buxton, 2010; Lee & Luykx, 2006). The result is that some women and students of color may feel less fit within a culture of science that centers on studying with and about people who do not look like them while also being taught by faculty members who are typically White men. Thus, many women and students of color feel themselves to be isolated outcasts in STEM fields (Bryant, 2019; Ireland et al., 2018; Ong et al., 2011).

Statement of the Problem

According to the current literature on African American women in STEM (Ireland et al., 2018), the lived experiences of African American girls and women in K-16 math and science classrooms remain insufficiently investigated. An understanding of these experiences may help to explain why some African American women at Town University, a PWI, continue to succeed in STEM majors despite obstacles whereas others switch to health science or social science majors. Few scholars have conducted research to understand the connection of African American girls' early exposure to science or math and their science or math identity formation (Pennock, 2009). Even fewer researchers have focused on the use of culturally responsive teaching strategies and caring instruction for STEM students (Gay, 2018; Siddle Walker, 2001) to engage African American girl learners in math and science classrooms (Farinde & Lewis, 2012; Young et al., 2017). There is also a limited understanding of how the hidden curriculum in STEM deters some African American women from pursuing STEM degrees. Moreover, there is a dearth of knowledge about the support networks for African American girls and women in STEM fields (Borum & Walker, 2011; Britt et al., 2010; Coneal, 2012; Fries-Britt & Holmes, 2012; Hanson, 2009; Ireland et al., 2018; Parker, 2013). Finally, there is scant literature on the cooling-out function in higher education, specifically at four-year institutions (Baird, 2014) that pushes African American women out of STEM majors.

Research Methods

This research is a qualitative study on the persistence of 16 African American women in STEM, social science, and health science programs at a PWI in the Midwestern part of the United States, denoted herein as Town University. The main question guiding this research study was "What are the lived experiences of African American women who remain in STEM majors and those who switch to social science and health science majors?" The subsidiary questions were as follows:

(1) Does early exposure to science or math influence African American girls' engagement in STEM?

(2) How do teachers influence African American girls' engagement in STEM?

(3) What are the science and math teaching methods that promote student engagement for African American girls?

(4) What is the hidden curriculum within STEM majors at Town University?

(5) How does social capital influence the persistence of African American girls and women in STEM?

(6) How does the cooling-out process within the STEM culture function to push African American women out of college STEM majors at a PWI?

This exploratory study employed the qualitative method of a multiple case study to obtain an in-depth understanding of the lived experiences of participants who persisted in

STEM, social science, and health science majors. A case study is defined as an empirical inquiry that investigates "a contemporary phenomenon within its real-life context, especially when the boundaries between the phenomenon and context are not clearly evident" (Yin, 2003, p. 13). The purpose of case studies is to gain insight into a process by studying in detail a specific case such as a person, situation, or problem (Krathwohl, 2004; Stake, 1995).

The multiple case study design involves replicating an experiment and comparing and contrasting findings (Yin, 2009). Multiple case studies also involve conducting research on the same phenomenon using multiple purposefully chosen cases to determine whether similar or different results are produced by studying multiple cases in the same context (Yin, 2003). Therefore, the multiple case research design is appropriate for this study to investigate whether there are differences in the lived experiences of African American women in STEM majors and those who switched from STEM majors to social science or health science majors in K–16 settings.

In this study, 16 African American women represent the multiple cases, which are bounded by school contexts (e.g., elementary, middle, high school, and college), because these contexts exposed the participants to science and math. These contexts also may have influenced their continuation in or departure from STEM majors. This study employed interviews with the same protocols to determine whether there were differences in the lived experiences, specifically within the school context, of African American women students who remained in or departed from STEM majors at Town University.

The multiple case study design also requires that participants be purposefully chosen. Purposive sampling is used to "identify the purposefully selected sites or individuals for the proposed study" (Creswell, 2003, p. 185). This research involved 16 undergraduate African American women who were purposefully chosen for inclusion in this study. Eight of the participants remained as STEM majors, whereas seven participants transferred from STEM majors into social science or health science majors. In addition, one study participant was in the process of switching from the health science major to chemistry, a STEM major.

The recruitment of these African American women took place by email through higher education administrators. The students voluntarily elected to participate in the study by emailing me and scheduling times to meet for the study beginning in March 2011 and ending in December 2011. Between May 2020 and July 2020, the former participants were contacted to obtain an update on their educational trajectories and careers by email survey or by interview (phone or Zoom).

The criteria for participation in the study were as follows: (1) women who self-identified as African American or Black, (2) students at Town University between the ages of 19 and 23, and (3) upperclassmen who were pursuing STEM majors or had transferred from STEM majors into health science or social science majors.

Following this chapter, readers will learn about the lived experiences of the sensational 16, African American women in STEM majors and those who switched from STEM majors to health science and social science majors. Chapter 1 focuses on early STEM success. In the next chapters 2 and 3, to better understand the lived experiences of the African American women, I draw upon the field of education to explain the teaching styles (e.g., culturally responsive teaching and caring instruction) that support their learning styles. I also employ the sociological frameworks of the hidden curriculum (chapter 4), social capital (chapter 5) and cooling-out (chapter 6) to understand their lived experiences in STEM majors. Chapter 7 focuses on the graduation and career aspirations of the African American women. Chapter 8 provides recommendations for STEM student success. Chapter 9 describes the conclusions and call to action for African American girls and women in STEM to keep on dreaming and pursuing undergraduate and graduate degree in STEM fields and medicine.

References

American Association of University Women. (1992). *How schools shortchange girls: A study of major findings on girls and education: The AAUW Report*. Washington, D.C.: American Association of University Women.
Baird, A. (2014). *The social function of for-profit higher education in the United States* [Doctoral dissertation, University of Central Florida]. UCF Electronic Theses and Dissertations.
Blalock, S. D., & Sharpe, R. V. (2012). You go girl! Trends in educational attainment of Black women. In C. R. Chambers & R. V. Sharpe (Eds.), *Black female undergraduates on campus: Successes and challenges (Diversity in Higher Education, Vol.12)* (pp. 1-41). Bingley, UK: Emerald Group Publishing Ltd.
Borum, V. & Walker, E. (2011). Why didn't I know? Black women mathematicians and their avenues of exposure to the doctorate. *Journal of Women and Minorities in Science and Engineering, 17*(4), 357-369. doi: 10.1615/JWomenMinorScienEng.v17.i4

Bowen, W. G., Chingos, M.M., & McPherson, M. S. (2009). *Crossing the finish line: Completing college at America's public universities*. Princeton: Princeton University Press.

Britt, S. L., Younger, T. K., & Hall, W. D. (2010). Lessons from high-achieving students of color in physics. *New Directions for Institutional Research, 148,* 75-83. https://doi.org/10.1002/ir.363

Brown, J. (2011). *African American women chemists*. New York: Oxford University Press.

Bryant, T. (2019). *Unhidden and unrelenting figures: The persistence of Black women in STEM disciplines* [Doctoral dissertation, California State University]. ProQuest Dissertation and Theses Global.

Bush, V. B., Chambers, C.R., & Wapole, M. (2009). Introduction. In V.B. Bush, C.R. Chambers, & M. Wapole (Eds.), *From diplomas to doctorates: The success of Black women in higher education and its implications for educational opportunities for all* (pp. 1-18). Sterling, VA: Stylus Publishing.

Committee on Barriers and Opportunities in Completing 2-Year and 4-Year STEM Degrees, Board on Science Education, Division of Behavioral and Social Sciences and Education, Board on Higher Education and Workforce, Policy and Global Affairs, National Academy of Engineering; National Academies of Sciences, Engineering, and Medicine. (2016). The culture of undergraduate STEM education. In S. Malcom & M Feder (Eds.), *Barriers and opportunities for 2-year and 4-year STEM degrees systemic change to support students' diverse pathways* (pp. 59-81). Washington, D.C.: National Academies Press. https://doi.org/10.17226/21739

Coneal, W. B. (2012). African American high-achieving girls: STEM careers as options. In C. R. Chambers & R. V. Sharpe (Eds.), *Black female undergraduates on campus: Successes and challenges* (Diversity in Higher Education, Vol.12) (pp. 161-183). Bingley, UK: Emerald Group Publishing Ltd.

Creswell, J. W. (2003). *Research design: Qualitative, quantitative, and mixed methods approaches.* Thousand Oakes: Sage Publications.

Dixson, A. & Chambers. C. R. (2009). College predisposition and the dilemma of being black and female in high school. In V.B. Bush, C.R. Chambers, & M. Wapole (Eds.), *From diplomas to doctorates: The success of Black women in higher education and its implications for educational opportunities for all* (pp. 21-38). Sterling, VA: Stylus Publishing.

Dortch, D. & Patel. C. (2017). Black undergraduate women and their sense of belonging in STEM at predominantly White institutions. *NASPA Journal About Women in Higher Education*, *10*(2), 202-215. https://doi.org/10.1080/19407882.2017.1331854

Evans, S. Y. (2007). *Black women in the ivory tower, 1850-1954: An intellectual history.* Gainesville: University Press of Florida.

Farinde, A.A., & Lewis, C. (2012). The underrepresentation of African American female students in stem fields: Implications for classroom teachers. *US-China Education Review*, 421-430.

Fries-Britt, S., & Holmes, K. (2012). Prepared and progressing: Black women in physics. In C. R. Chambers & R. V. Sharpe (Eds.), *Black female undergraduates on campus: Successes and challenges (Diversity in Higher Education, Vol. 12)* (pp. 199-218). Bingley, UK: Emerald Group Publishing Ltd.

Gasman, M., Smith, T., Ye, C., & Nguyen, T. (2017). HBCUs and the production of doctors. *AIMS Public Health*, *4*(6), 579-589. doi: 10.3934/publichealth.2017.6.579

Gay, G. (2018). *Culturally responsive teaching: Theory, research, and practice*. New York: Teachers College Press.

Gold, S.P. (2011), Buried treasure: Community cultural wealth among Black American female students. In C. R. Chambers (Ed.), *Support systems and services for diverse populations: Considering the intersection of race, gender, and the needs of Black female undergraduates (Diversity in Higher Education, Volume 8)* (pp. 59-72). Bingley, UK: Emerald Group Publishing Limited.

Hanson, S.L. (2009). *Swimming against the tide: African American girls and science education*. Philadelphia: Temple University Press.

Harding, S. (2006). *Science and social inequality: Feminist and poststructural issues*. Urbana: University of Illinois Press.

Hill, C., Corbett, C., & St. Rose, A. (2010). *Why so few?: Women in science, technology, engineering, and mathematics*. Washington, D.C.: American Association of University Women.

Ireland, D. T., Freeman, K.E., Winston-Proctor, C.E., DeLaine, K. D., Lowe, S. M., & Woodson, K. M. (2018). (Un)hidden figures: A synthesis of research examining the intersectional experiences of Black women and girls in stem education. *Review of Research in Education, 42*, 226–254. https://doi.org/10.3102/0091732X18759072

Jez, S. J. (2012). Analyzing the female advantage in college access among African Americans. In C. R. Chambers & R. V. Sharpe (Eds.), *Black female undergraduates on campus: Successes and challenges* (Diversity in Higher Education, Vol.12) (pp. 43- 57). Bingley, UK: Emerald Group Publishing Ltd.

Jordan, D. (2006). *Sisters in science: Conversations with Black women scientists on race, gender, and their passion for science*. West Lafayette: Purdue University.

Krathwohl, D. R. (2004). *Methods of educational and social science research: An integrated approach*. Long Grove: Waveland Press.

Lee, O., & Buxton, C. A. (2010). *Diversity and equity in science education: Research, policy, and practice*. New York: Teachers College Press.

Lee, O., & Luykx, A. (2006). *Science education and student diversity*. New York: Cambridge University Press.

Malcom, L., & Malcom, S. (2011). The double bind: The next generation. *Harvard Educational Review, 81*(2), 162-172. https://doi.org/10.17763/haer.81.2.a84201x508406327

Malcom, S., Hall, P., & Brown, J. (1976). *The double bind: The price of being a minority woman in science*. Washington, DC: American Association for the Advancement of Science.

McPherson, E. (2013). Each one, teach one: Black women's historical contributions to education. In D.J. Davis and C. Chaney (Eds.), *Black women in leadership their historical and contemporary contributions* (pp. 55-68). New York: Peter Lang.

National Science Foundation. (2019). *Minorities, women, and persons with disabilities in science and engineering: 2019* [Data set]. https://ncses.nsf.gov/pubs/nsf19304/data

Ong, M., Wright, C., Espinosa, L. L., & Orfield, G. (2011). Inside the double bind: A synthesis of empirical research on undergraduate and graduate women of color in science, technology, engineering, and mathematics. *Harvard Educational Review, 81*(2), 172-208. https://doi.org/10.17763/haer.81.2.t022245n7x4752v2

Parker, A. D. (2013). *Family matters familial support and science identity formation for African American female scholars* [Unpublished doctoral dissertation]. Charlotte: University of North Carolina-Charlotte.

Pennock, P. H. (2009). *African-American girls and scientific argumentation: Lived experiences, intersecting identities and their roles in constructing and evaluating claims* [Doctoral dissertation, Western Michigan University]. ScholarWorks@WMU.

Perna, L., Lundy-Wagner, V., Drezner, N.D., Gasman, M., Yoon, S., Bose, E., & Gary, S. (2009). The contribution of HBCUs to the preparation of African American women for STEM careers: A case study. *Research in Higher Education, 50,* 1-23. doi: 10.1007/s11162-008-9110-y

Robinson, S., & Franklin, V. (2011). Working against the odds: The undergraduate support needs of African American women. In C. R. Chambers (Ed.), *Support systems and services for diverse populations: Considering the intersection of race, gender, and the needs of Black female undergraduates (Diversity in Higher Education, Volume 8)* (pp. 21-41). Bingley, UK: Emerald Group Publishing Limited.

Siddle-Walker, V. (2001). African American teaching in the south: 1940-1960. *American Educational Research Journal, 38*(4),751-779. https://www.jstor.org/stable/3202502

Stake, R. E. (1995). The art of case study research. Thousand Oaks, CA: Sage Publications.

U.S. Department of Education & National Center for Education Statistics. (2019). *Status and Trends in the Education of Racial and Ethnic Groups 2018* (NCES 2019-038), *Degrees Awarded.* https://nces.ed.gov/pubs2019/2019038.pdf

Wapole, M. (2009). African American women at highly selective colleges: How African American campus communities shape experiences. In V.B. Bush, C.R. Chambers, & M. Wapole (Eds.), *From diplomas to doctorates: The success of Black women in higher education and its implications for educational opportunities for all* (pp. 85-107). Sterling, VA: Stylus Publishing.

Warren, W. (1999). *Black women scientists in the United States*. Bloomington: Indiana University Press.

Yin, R. K. (2003). *Case study research: Design and methods*. Thousand Oaks: Sage Publications.

Yin, R. K. (2009). How to do better case studies (with illustrations from 20 exemplary case studies). In L. Bickman & D.J. Rog (Eds.), *The sage handbook of applied social research methods* (pp. 254-282). Los Angeles: Sage Publications.

Young, J. L., Feille, K. K., & Young, J. R. (2017). Black girls as learners and doers of science: A single-group summary of elementary science achievement. *Journal of Science Education, 21*(2) 1-20.

Chapter 1

African American Girls' Experiences in Science and Math Classrooms in K-12 settings

This chapter addresses the question, "Does early exposure to science or math influence African American girls' engagement in science, technology, engineering, and math (STEM)?" In elementary school, teachers actively choose to use projects as a method of introducing African American girls to science and math projects. This engagement builds the foundation for their later engagement in STEM. Many of these students also experience early success in math and science in K-12 settings when learning the course materials or teaching the course materials to their peers.

Science Projects

Within science classrooms, one way in which teachers expose African American girls to science is through projects. For instance, several of the study participants recounted early exposure to science projects in school. For example, Rachelle, a biology major, remembered mini science projects as early as preschool. The earliest memory of science for Danielle, a health science major and former biology major, was a project on conduction. She also recalled a number of dissections, including a chicken wing and frogs as well as a lot of chemical studies. Briana, another biology major, elaborated further:

> I was in the gifted science program at my elementary school. My first memories of science were when we did little things like paper chromatography, measuring

the heart rate, because it was to prepare us for a field trip that we were taking. We were going to pretend that we were on spaceships. We were in outer space, and everybody had a certain role.

Similarly, Celeste, a sociology major and former pre-business major, reported that her science experiences in grade school involved a lot of hands-on projects rather than reading out of a textbook. Specific projects she recalled included making lamps out of pop cans, building model rockets from scratch, and constructing a bomb. For Amber, a health science major and former biology major, memorable projects included studying beehives and electrical circuits. She explained, "the big projects, for me, probably were the big successes because, when I did a project, I always went above and beyond. It gave me a chance to be creative with projects."

These individual accounts suggest that teachers play an important role in presenting science to African American girls and that the use of projects constitutes a memorable part of this early introduction to science. This exposure to science helps to develop the science identities of African American girls who want to pursue science as a major in college or as a career (Ireland et al., 2018; Smith, 2016). Science identity formation is "a process that people go through to identify as a scientist" (McPherson, 2012, p. 5). Science identity formation centers on competence in science; demonstrated performance of scientific activities; and recognition by oneself and others, as part of the scientific community (Carlone & Johnson, 2007).

Math Projects

Similar to science teachers, math teachers often present their subject to African American girls through math projects. Three study participants described their early experiences with math within the classroom. For Ashley, a biology major, the earliest memories were from second grade. She remarked:

> At the start of class, my teacher would give us four or five math problems to figure out. I would go through them really fast. I always wanted to be the first one done, because if you were the first person done and you got them all correct, then you were able to go around and grade everybody's paper. I wanted to be that kid. In elementary school all the way up to eighth grade, I was really good at math. I actually liked math more than I liked science, although I was also really good at science.

Likewise, Jennifer, a math major, remembered that in elementary school, "my class had a problem of the week every week that I would do again with my dad. I distinctly remember doing that. With each year, in my elementary school, I remembered what I did with my math classes." Another example of a memorable math project in elementary school was provided by Patricia, a business major. She reported, "For one of my classes, I think that it was fourth grade, we had a little business in which we were basically selling candy before and after school." Projects of this type allow students such as Patricia and her classmates to practice their math skills in a context familiar to their everyday lives.

These stories illustrate that early exposure to math through math projects assigned by their teachers can play a pivotal role in the development of the math identities of African American girls. Consistent with prior literature (Ireland et al., 2018; Smith, 2016), this study found that African American girls develop math identities based on their experiences in K–12 schools as well.

Early Success in Science and Math

The participants of this study reported feeling successful in science and math in middle school during their African American girlhood. For instance, Ashley, a biology major, explained, "I felt most successful in science when I was a part of the academic team. I was included on the team specifically for being able to answer the science questions during the competition." Celeste, a sociology major and former pre-business major, also related her feelings of success to her experience in a science competition. She stated, "I felt smart when I won the science fair in the 7th grade because that was the first project I ever did, and I didn't know what I was doing, but we won." Raven, a current health science major and former biology major, described her feelings of success as follows: "my successes were the first experience of working in the lab and seeing how things worked when I mixed different chemicals or random objects that you see at home to see how they interact." Another biology major, Briana, implied feelings of success through the enthusiasm with which she described her engagement with science class. She stated:

> I did love science class in eighth grade, because we were learning elements. I thought that the periodic table was the coolest thing in the world. We did a forensics project and a crime scene investigation. We had to figure out who killed the person. We did fingerprinting too.

In a similar fashion, several study participants shared their math successes from middle school during their African American girlhood. For instance, Regina, a sociology major and former pre-business major, stated that being successful centered on working in teams. She said "just working with other students to figure out the answer. I have a tendency to work by myself. So, working with other students to achieve the answer was successful, because I could get into math." Another example of a math success story was that of Simone, an engineering major. She said that she felt successful "when I started doing algebra and I knew people in high school who were doing it." Jennifer, a math major at Town University, described her experience in math as follows:

> I was always getting put in the top tiers, the accelerated classes. And then I was always doing the best in those classes. All of my math teachers in middle school always loved me; they would always pull me aside and tell me, 'You did the best on this test.' So, that was really nice.

For Patricia, a business major, feelings of success were also fostered by her math teacher. She explained:

> In math, I know that I was always recognized by my teacher, because she knew that I was at the top and that I knew what I was doing…She would always ask, 'Can you help the rest of the students?'

Finally, when asked about her experience of success in math, math major Lara provided as evidence "my grades and the fact that people would always come to me for help."

These narratives show how African American girls experienced early success in science and math, because they mastered the materials and some even taught the materials to their peers. These successes also contributed to their science and math identity development in their adolescent years.

Conclusion

This chapter has described how African American women are exposed to science and math early in their educations through science and math projects during girlhood. Since teachers play a significant role in the math and science identity development of African American girls, it is critical that they have access to resources (e.g., books and materials) for science and math projects. In addition, in adolescence, African American girls have early successes by learning the materials; engaging in projects; and in some cases, teaching science or math to their peers. It is thus important for teachers to provide African American girls with leadership opportunities in the classroom, opportunities to engage in hands-on learning projects, and recognition when they are successful within math and science classrooms. Thus, during teacher training, there is a critical need to provide tools to help teachers further develop the math and science identities of African American girls throughout middle school and high school.

References

Carlone, H. B., & Johnson, A. (2007). Understanding the science experiences of successful women of color: science identity as an analytic lens. *Journal of Research in Science Teaching, 44*(8), 1187-1218. https://doi.org/10.1002/tea.20237

Ireland, D. T., Freeman, K.E., Winston-Proctor, C.E., DeLaine, K. D., Lowe, S. M., & Woodson, K. M. (2018). (Un)hidden figures: A synthesis of research examining the intersectional experiences of Black women and girls in stem education. *Review of Research in Education, 42*, 226–254. https://doi.org/10.3102/0091732X18759072

McPherson, E. (2012). *Undergraduate African American women's narratives on persistence in science majors* [Doctoral dissertation, University of Illinois at Urbana-Champaign]. Illinois Digital Environment for Access to Learning and Scholarship.

Smith, M.R. (2016). *Black pearls: Examining the science identity development of African American girls in a culturally relevant STEM counterspace* [Doctoral dissertation. University of California-Los Angeles]. UCLA Electronic Theses and Dissertations.

Chapter 2

African American Girls' Student Engagement in Science and Math in K-12 Schools

Chapter 1 explained the important role that teachers play in introducing African American girls to science and math. This chapter asks the question, "How do teachers influence African American girls' engagement in science, technology, engineering, and math (STEM)?" The chapter explores the effect of teachers on the engagement of African American girls in science and math classes in K-12 schooling through the use of culturally responsive teaching. The discussion considers culturally responsive teaching by African American girls' leadership in groups, followed by their class participation. It continues by describing the use of culturally responsive teaching in these students' engagement in science projects, as well as their involvement in enriched curriculum.

Leadership in the Classroom

As early as elementary school, African American girls exhibit leadership skills within the classroom. For example, Ashley, a biology major, explained:

> A lot of times in math class, we would be given problem sets, and I would be the one going up to the board and explaining, working it out on the board and explaining why it worked and how you did it.

Similarly, Amber, a health science major, reported, "naturally, even if I didn't want to be the group leader, I've always been one." Further elaborating on her leadership style, she said, "I tried to incorporate everybody's ideas." Raven, another health science major, also found herself taking a lead in math class as well. She described her experience as follows:

> Oftentimes, other students would ask me for help in math, because they knew that I would get it. They would say, "Since you understand it, can you explain it?" So, I guess you can say that I was a designated team leader.

These passages show that African American girls' leadership activities in elementary school classrooms include showing classmates how to do problem sets and helping peers with course assignments by answering questions.

In middle and high school, African American girls continue to showcase their leadership skills within the classroom through group work projects and by helping their peers. For instance, Kayla, a chemistry major and former health science major, expressed her experience as follows: "Most of the time I did take the lead in group work during middle school. I feel like my peers looked up to me when they were confused, so I kind of took that as a responsibility." Patricia, a business major, also provided leadership among her peers in middle school. She stated:

> I was looked at as a leader in some of my classes such as math. Other students would consider me as this "smart girl", so people would look to me and expect me to be the leader. Also, I liked computers. If we had group projects in that class, of course I would step up and lead.

In describing her high school experience, Jennifer, a math major, reported that she engaged in group leadership a little bit. She further elaborated:

> There were several times in high school when I was in a group and there were too many people trying to be the leader. So, in those cases, I backed down. There were a few times when I consciously decided not to be the leader.

These narratives demonstrate that African American girls' leadership in the classroom persisted in middle and high school.

Course Participation

In elementary school, the African American women in this study were comfortable participating in their science and math classes during girlhood. This is exemplified by the narratives of several study participants. For instance, Crystal, a health science major, explained that she participated in math class because "it was interesting. I liked to go to the board and write math problems out." Similarly, Amber, a health science major, reported that she participated in science class because it was fun. She stated:

> Science was fun. The way we learned in elementary school about science with hands-on projects was a lot of fun. You did it, and you learned, and it was a great thing, and it was fun. You learned at your own pace, and you got to do and experience it at your own pace. I will be honest; school didn't feel like "oppression" until college.

Another way in which class participation was demonstrated is through African American women's willingness to ask questions in math and science courses during girlhood. For example, Ashley, a biology major, simply stated, "if I wanted to know something in science class, then I would ask." Briana, also a biology major, detailed her strategy for asking questions in math. She explained:

> I did not really have a lot of questions in my science classes, but my math classes were much more difficult, so I would ask more questions in those classes. I do remember that my math teacher was helpful even though I did not really like him because he gave out so much homework. He did explain things better when you asked him outside of class or asked him during class about a certain problem. We would always have time to do homework in class. It was during those times, when the whole class was dead silent that I would go up and ask him questions. I also remember doing after-school tutoring a couple of times. That involved him just showing me how to do a problem, then me sitting and trying to work out the problems and then saying, "Okay, can you help me out with this?"

Similarly, Carmen, a sociology major and former biology major, felt confident of her abilities in science and asked questions as a means of furthering her knowledge when necessary. Carmen reported:

> Overall, school was a time for me to be the shining star in science. If I had a question, I would ask it, although I do not think that I had many questions at that time. I am still the kind of person who, if I do have a question, I learn better if I try to figure it out myself. But if it is something that I cannot get, then I will ask.

These biographical accounts reveal that the African American women in this study participated in their math and science classes in elementary school. The classroom structure also allowed for them to pose questions and get their questions answered. They also participated by going to the board to solve problems. Gay (2018) defined culturally responsive teaching as "using the cultural knowledge, prior experiences, frames of reference, and performance styles of ethnically diverse students to make learning encounters more relevant and effective for them" (p. 36). Moreover, as shown in prior research on teaching and learning in elementary and middle school (King, 2017), African American girls are engaged when culturally responsive teaching allows for them to own their learning. Thus, from the narratives of the African American women study participants, one can conclude that they experienced culturally responsive teaching in primary school.

Similar to elementary school, the African American women in this study experienced culturally responsive teaching in middle school that allowed them to actively engage in science and math classes. The study participants continued to be engaged in their science and math classes as evidenced by their asking questions of their instructors and providing answers to questions asked by the teacher. Regina, a sociology major and former pre-business major, engaged in both of these types of participation. She explained:

> In science, if I happened to know an answer, then I participated. In math, I probably did a bit more asking questions in class. When the teacher asked, 'Is everyone clear?' that was when I would raise my hand. I was more trying to figure out the materials, not just trying to answer.

Similarly, Jennifer, a math major, said, "I participated in math because I liked to ask questions." For several other study participants, participation involved answering questions more than asking them. As stated by Danielle, a health science major, "Math was something that I understood. It was something that I was comfortable with and knowledgeable about." Shannon, a biology major, described a similar type of participation. She remarked: "In science, I participated because I knew the answers. Or if I did not know the answers, then I would want to know the answers." Carmen, a sociology major and former biology major, elaborated further:

> By my eighth grade science class, I had come into my own, and I was confident. I felt like "I know the answer. And so, you need to know that I know the answer and I am better than you." That was when I would start to answer questions.

These anecdotes are consistent with the middle school literature on culturally relevant pedagogy, which allows students to actively engage in science and math content (Laughter & Adams, 2012).

The African American women in this study also benefited from culturally responsive teaching in their science and math courses in high school. They participated in class to learn concepts and to show their knowledge of the course materials. For instance, A math major, Lara, stated:

> I participated in science because, I knew it. For math, I participated often when we had stuff to participate in. We kind of worked in groups. There were some days when we did stuff on the board. Everybody had to participate at some point, so, I just did.

In a similar fashion, Crystal, a health science major, described her participation in science class as follows: "In science, I participated just to see if I was getting the materials or to see if I was understanding what was going on correctly." In a comparable manor, Kayla, a chemistry major and former health science major, reported, "In science, I participated mostly to ask questions to grasp the concepts. In math lessons, I was always at the board." Briana, a biology major, also participated in various ways in her science and math classes. She said:

> Since I was engaged and I was interested in the science courses, I was pretty willing to participate. I was engaged in the class and the topics that we were talking about. It made sense. For math, a lot of the time, I was lost. It was a lot harder to pick up the concepts in math than it was in science. I would constantly be asking questions and trying to make sure that I could grasp the concepts enough to solve the problems in our group. I raised my hand all of the time in class in front of all of my peers.

Finally, Jennifer, a math major, stated:

> For math, all my teachers actually greatly appreciated that I asked a lot of questions. The teachers were always happy that I asked questions because a lot of people in the classes didn't and it just helped everyone learn and go further.

These stories thus point to the fact that African American women in this study participated in math and science because they were interested in the course materials. They also asked questions to make sure that they understood the content and to ensure that they would be able to complete problem sets.

Science Projects

African American girls were very engaged in science projects in K-12 schooling. For instance, Amber, a health science major, recalled a girlhood experience from elementary school. She recollected:

> My first memory would be in fourth grade. We had to do a class project with electricity, and I'm the type of person who likes to go above and beyond. We had to build our own circuit board. We either had to make a light bulb work or a bell, and I made both work at the same time, so I was an overachiever.

In a similar fashion, Briana, a biology major, described a science project experience from middle school. She stated: "I remember that in sixth grade we built rockets. We got our own rocket kits, and we spray painted them." Carmen, a sociology major, and former biology major, recalled a science experiment in middle school. She explained:

> I remember a Cheeto experiment, where they lit the Cheeto, put the Cheeto in acid, and then put it in helium. And we dropped it in a blue thing and watched it explode. You did not get to touch anything as students. The teachers were not going to be liable for children's fingers blowing up.

Experience with science projects also continued in high school, as recounted by Patricia, a business major. She said:

> In high school, I liked the research projects, because they allowed us to research the outside. I think that we were working on climate. That was interesting. I liked the dissecting part. I don't like touching animals, but it was interesting to see.

Crystal, a health science major, also told a story of completing science projects in high school. She explained:

> We did a lot of experiments and dissecting. I think that I probably only had to do more projects in high school. It was more hands-on. We had pigs, cats, worms, and frogs. We had little packs and utensils to do the dissections. For chemistry, it was a lot of math-related stuff. We also had chemicals in chemistry to do different combustions.

Simone, an engineering major, shared her experience of completing a science project in high school as well. She stated: "The only thing that comes to my mind is when I learned how to make a volcano and when I took the shell off of an egg. I think that I soaked it in vinegar."

This section emphasizes the importance of engaging African American girls in hands-on learning not just during elementary school but also during adolescence to foster their interests in science. The study participants' responses also support the use of culturally responsive teaching practices in middle school and high school settings through techniques such as hands-on learning (Farinde & Lewis, 2012).

Enriched Curriculum

As illustrated in the preceding section, the African American women in this study participated in high school curricula that exposed them to advanced science and math curricula. In fact, the majority of these women took advanced placement (AP) and honors curricula for math or science or both. Table 1 summarizes the curricula reported by the study participants.

Table 1

High School Math and Science Curricula Reported by Study Participants

Name	Major	Science Courses	Math Courses
Danielle	Health science, formerly biology	Anatomy AP Biology Honors Biology Chemistry Honors Physics	Algebra Geometry
Raven	Health science, formerly biology	Anatomy AP Biology Biology AP Chemistry Chemistry Physiology	Algebra Geometry Trigonometry AP Statistics
Crystal	Health science, formerly biology	Biology Chemistry	Geometry Trigonometry Precalculus AP Economics
Amber	Health science, formerly biology	Biology AP Biology Chemistry Physics	Algebra II Precalculus
Regina	Sociology, formerly pre-business	Biology Chemistry Lab Science Advanced Laboratory Science	Algebra, Advanced Algebra Geometry
Carmen	Sociology, formerly biology	Honors Biology Chemistry Physics	Honors Algebra Honors Precalculus Honors Calculus
Celeste	Sociology, formerly pre-business	Biology Chemistry Physical Science	Pre-Algebra Algebra II Advanced Math Geometry

Table 1 Continued

High School Math and Science Curricula Reported by Study Participants

Name	Major	Science Courses	Math Courses
Kayla	Chemistry, formerly health science	AP Chemistry Integrated Science I Integrated Science II	Algebra I Geometry Algebra II/Trigonometry Precalculus
Briana	Biology	Biology Chemistry AP Chemistry Physics	Geometry Algebra II Trigonometry AP Calculus
Rachelle	Biology	Anatomy Biology Chemistry Physics	Algebra I Algebra II Precalculus Trigonometry
Shannon	Biology	Biology Advanced Biology Chemistry Physics	College Algebra Honors Geometry Precalculus Honors Trigonometry AP Statistics
Ashley	Biology	Aquatic Science Biology AP Biology Chemistry	Algebra Geometry Trigonometry Precalculus
Patricia	Business	Honors Biology Honors Chemistry Honors Physical Science AP Physics	Algebra Geometry Precalculus AP Calculus
Simone	Engineering	Biology Chemistry	Algebra II Geometry Trigonometry Precalculus Calculus AB and BC Honors Statistics

Table 1 Continued

High School Math and Science Curricula Reported by Study Participants

Name	Major	Science Courses	Math Courses
Jennifer	Math	Biology Biology II AP Biology Chemistry AP Chemistry AP Chemistry II Physics Accelerated Physics	Algebra Algebra II Precalculus
Lara	Math	Biology Chemistry AP Chemistry	Geometry Algebra/Trigonometry Precalculus AP Calculus Financial Math

The data in table 1 demonstrates that the African American women in STEM majors at Town University, a predominantly White institution (PWI) were engaged in advanced science and math curricula in high school, often taking honors and advanced placement (AP) courses. Consistent with prior research (Thiry, 2019), this type of curriculum should have prepared them for science and math studies in college.

Conclusion

This chapter has demonstrated the positive effect that the use of culturally responsive teaching in K-12 schooling has on African American girls in science and math classrooms. Specifically, with culturally responsive teaching, African American girls take on leadership roles in science and math classrooms by helping peers during group work or solving problems in front of the class. As a result of this culturally responsive teaching, African American girls have a safe space to learn by asking questions or answering instructors' questions. They also have access to enriched curricula, such as AP and honors classes to prepare for the rigors of STEM coursework in college.

References

Farinde, A. A., & Lewis, C. (2012). The underrepresentation of African American female students in stem fields: Implications for classroom teachers. *US-China Education Review*, 421-430.

Gay, G. (2018). *Culturally responsive teaching: Theory, research, and practice*. New York: Teachers College Press.

King, N. S. (2017). When teachers get it right: Voices of Black girls' informal STEM learning experiences. *Journal of Multicultural Affairs*, 2(1), 1-15. https://scholarworks.sfasu.edu/jma/vol2/iss1/5

Laughter, J. C., & Adams, A. D. (2012). Culturally relevant science teaching in middle school. *Urban Education*, 47(6),1106-1134. https://doi.org/10.1177/0042085912454443

Thiry, H. (2019). Issues with high school preparation and transition to college. In E. Seymour & A. Hunter (Eds.), *Talking about leaving revisited: Persistence, relocation, and loss in undergraduate education* (pp. 137-147). Switzerland: Springer.

Chapter 3

Teaching and Learning Science and Math in K-12 School Settings

Chapter 3 considers the question "What are the science and math teaching methods that promote student engagement for African American girls?" This chapter discusses how African American girls learn in science and math classrooms despite the lack of racial diversity in teacher demographics. In K-12 learning environments, caring instruction and culturally responsive teaching provided by teachers enable African American girls to learn science and math. Teachers also use culturally responsive teaching, including hands-on learning, and caring instruction through teaching using a step-by-step approach to make sure that African American girls grasp the science and math course materials. Such teaching methods promote student engagement among African American girls within this context. This chapter ends by presenting the study participants' reflections on their preparation for college-level courses in science and math at Town University.

K-12 Learning Environment in Science and Math

In their elementary school science and math classes, African American women in this study encountered a warm learning environment with their peers in girlhood. A warm learning environment is one in which teachers hold high expectations for the success of the learners and commit to helping students be successful and students are helpful to their peers (Bondy et al., 2013). When asked about relationships with peers, Regina, Rachelle, and Ashley spoke about being with friends and/or peers inside warm classrooms. Regina, sociology major, and former pre-business major said:

> I would say my closest friends—for some reason, we all had the same math class. Then, there was that one student who was the genius of the class, we would occasionally work with him. I would say all of the students really worked together.

Similarly, Rachelle, a biology major, stated "For math, I guess more of my female friends, because we were closer outside of class and we worked better in class as well." Further, Ashley, a biology major, explained her relationship with her classmates as follows: "I feel that I have always worked well with anyone who is willing to work with me, regardless of gender, race, or class. I wanted to make sure that everyone was able to contribute to what we were doing."

In a similar fashion, Jennifer, a math major; Raven, health science major and former biology major; and Shannon, a biology major, all described working in groups or pairs in their math or science classes during middle school. For example, math major, Jennifer said:

> In math, I remember we'd do group work. I only remember that because I remember that my math class you sat at tables in sixth grade and my seventh grade year. My eighth grade year it was individual desks, but my sixth grade and seventh grade years you sat at either tables or in clusters. So, it was geared towards when it got time to do work, you worked with the people you were with and close to. Then if you ever did little competitions, you worked with them. My eighth

grade class, I remember we all sat individually and not in desks.

Raven, a health science major and former biology major, also reported working with other students. She explained: "in the science labs, we had to work together, because there just wasn't enough equipment for everyone to work by themselves. I believe that we did work well together. I am still trying to piece that together." Shannon, a biology major added, "I did enjoy my science experiences in middle school. I liked my teachers. I liked that we were doing more hands-on work. When we did labs, we worked with partners. Other than that, it was mostly independent studying."

In high school, the African American women continued to have solid relationships with their peers and friends with whom they completed science or math projects within the classroom. Briana, biology major stated: "in science classes, it was just basically, you and your partner would do the lab. You know most of the time my partner was a friend of mine or somebody I knew." Similarly, Danielle, health science major and former biology major, said that she worked with "peers. We always had fun whoever was sitting at my table. We got along with each other." Further Ashley, a biology major, pointed out:

> I worked well with pretty much everyone, but I do remember that during junior year in my AP [advanced placement] biology class, I worked in a group that consisted of myself, one of my friends who was also on the basketball team who was Hispanic, and two other African American girls who were in the class.

Likewise, math major Jennifer recalled, "I worked well with all my peers. I didn't have problems working with any of them. Like I said, because it started to be that you saw the same people." In contrast, business major Patricia was more selective about working with her peers. She explained:

In math, the only reason why I probably wouldn't work with others or would be skeptical or would strategically choose whom to work with was because people wanted me in their group because I was the "smart girl". I did not want people to take advantage of me or copy my answers or think that I would do all of the group work. So, I would try to work with people who would help me out and who would not leave me stuck with having to do work on my own.

These responses demonstrate that in girlhood, most of the African American women in this study were comfortable in their warm science and math classrooms in primary and secondary schools, because they built relationships with their peers, including working with them on assignments and projects. These warm environments also promoted their engagement in science and math. This type of environment is consistent with caring instruction. Siddle Walker (2001) posits that interpersonal caring in the classroom consists of building relationships between teachers and students inside and outside of the classroom, teachers holding high expectations of students, and teachers tailoring their teaching methods to help students learn class materials.

Teacher Demographics

Among the high school science and math teachers of the African American women, there was a lack of racial diversity. Some of the women did experience teaching by a few teachers of color, along with a number of White teachers. Others encountered only White instructors for all of their math and science classes in high school. Table 2 provides a summary of these teacher demographics as reported by the study participants.

Table 2

Demographics of the High School Math and Science Teachers as Reported by Study Participants

Name	Major	Science Instructors	Math Instructors
Danielle	Health science, formerly biology	White male	One White woman
Raven	Health science, formerly biology	African American (Biology instructor and other science instructors)	All African American women
			White man (statistics instructor)
		White and Arab (Chemistry instructors)	
Crystal	Health science, formerly biology	All White	All White
Amber	Health science, formerly biology	African American woman (Chemistry instructor)	All white
		White woman (Biology instructor)	
		White men (AP Biology and Physics instructors)	

Table 2 Continued

Demographics of the High School Math and Science Teachers as Reported by Study Participants

Name	Major	Science Instructors	Math Instructors
Regina	Sociology, formerly pre-business	White women	White men and White women
Carmen	Sociology, formerly biology	Two White women	All White women
Celeste	Sociology, formerly pre-business	White men (biology and chemistry instructors) White woman (physical science instructor)	African American woman (Pre-Algebra, Geometry, and Advanced math instructors) White man (Algebra II instructor)
Kayla	Chemistry, formerly health science	Two White women and one White man	White Men (math instructors)
Simone	Engineering	All White men (science instructors)	One White woman and two White men
Briana	Biology	Three White men and one White woman	Two white women and one white man
Rachelle	Biology	Majority White men White woman (Physics instructor)	One African American and majority White men and women

Table 2 Continued

Demographics of the High School Math and Science Teachers as Reported by Study Participants

Name	Major	Science Instructors	Math Instructors
Shannon	Biology	African American man (Chemistry)	African American woman (Geometry and Precalculus)
		White men (Biology, AP Biology, and Physics)	White men (College Algebra and AP Calculus)
			White woman (AP Statistics)
Ashley	Biology	Latina (Biology) White men (Aquatics, Chemistry)	African American woman (precalculus) White women (Algebra, Geometry, and Trigonometry)
Patricia	Business	One African American, two White women, and one White man	One White man and three White women
Lara	Math	African American woman and a White man	Two African American women, one Black man, and White man
Jennifer	Math	All men	Some women

The data in table 2 indicates that the women in this study received the majority of their science and math instruction in high school from White instructors, primarily men.

Culturally Responsive Teaching and Caring Instruction

In elementary school, African American women in this study received culturally responsive teaching from instructors in warm class environments that promoted student engagement, hands-on learning, and caring instruction. Culturally responsive teaching is defined as "using the cultural knowledge, prior experiences, frames of reference, and performance styles of ethnically diverse students to make learning encounters more relevant and effective to them" (Gay, 2018, p. 36). Culturally responsive teaching also provides caring instruction for diverse learners within a classroom. Culturally responsive teaching empowers students to be successful learners and transforms the lives of students of color by showing respect and making sure that they can be academically successful. Culturally responsive teaching is humanistic, or person-centered, in that it helps students learn about their culture and other cultures.

In addition to culturally responsive teaching, the ethic of care showed up in the K-12 math and science classrooms of the African American women. According to Howard (2001), "care, as an ethic in teaching, includes explicitly showing affective and nurturing behavior toward students, which can have a positive influence on student desire to learn" (p. 138). In this study, many participants reflected on the caring instruction that they received throughout their K-12 schooling and how it promoted student engagement and hands-on learning. For example, Rachelle, a biology major, provided examples of ways in which her elementary school teachers provided caring instruction. She pointed out:

> The teachers found a way to relate to everyone. If we had one student who was dragging, then the entire class was behind. They would stop the entire class

just to make sure that one person caught up. They kept it relevant to the entire class, and everything was clear. They cared about us as students, not only about our grades. If they saw something was going on outside of class, then they would reach out to you. They felt like second parents. A caring instructor is someone who reaches out to me outside of the classroom not only telling me about this homework is due but asking me "how did the homework go last night?" and "what do you have planned today?" They have an interest in you as a person not only as their student.

Reflecting on her elementary school experience, Danielle, a health science major and former biology major, also described caring instruction. She stated:

> My elementary school teacher was knowledgeable. She made it interesting for the class. She would walk around to see if we needed help with anything. She was very personable. She cared about our class. She wanted to be involved in our lives. She taught well for visual learners and kinesthetic learners. If we were better at hands-on stuff, then we got our opportunity to do that. If we were better at watching things happen, then we got our opportunity to do that. She kept it open for everyone to learn in their certain ways.

In a similar fashion, Regina, sociology major and former pre-business major, described caring instructors. She stated: "My math instructors were very friendly, very welcoming, very helpful. They knew the material. They knew how to teach it in an explanatory way so that students at our level could understand it. My science instructors tried to engage students." Crystal, a health science major and former biology major, elaborated on how her science teachers also tried to engage students in their classrooms. She pointed out:

> For science, it was hands-on. As they were writing on the board you were writing or trying to figure something out. They lectured very little, which was important when you were that young. You lose attention very fast. I think that was what I liked the most was the hands-on portion of things.

Kayla, a chemistry major and former health science major, further described caring instructors in elementary school. She explained:

> They were really motivating and dedicated to what they do. They put in a lot of effort. They would stay after school. Even during school, they would miss their own lunch to help students get more acquainted with the school.

Finally, biology major Shannon described the caring instruction and culturally responsive instruction in her lived experience. She stated:

> I really liked when my teachers would teach the lesson; ask for participation; and then, once they had taught it, sometimes to see how well we grasped the information, they would have little competitions at the board and you could go up and they would ask a question and they would see who could write the answer the quickest. I liked when they would do it like that. This happened every other day when they were teaching new materials.

The preceding comments support the idea that caring instruction and culturally responsive teaching within science and math classrooms involves the teacher engaging students in the course materials, using hands-on learning, being helpful, and making the course materials understandable for learners. Moreover, in such environments, the teachers are relatable and feel like an extension of family members.

Similarly, the study participants also experienced culturally responsive teaching and caring instruction in middle school. For instance, Briana, a biology major, mentioned one of her middle school teachers as a good teacher before identifying the qualities that she considers essential to good teaching. She pointed out:

> For science classes, my eighth-grade teacher was a good teacher and very knowledgeable. She was very hands-on, very helpful. A good teacher is very open to students' questions, very personable, very willing to take time to explain to you the concepts no matter how many times you ask or how many times you just do not understand. A good teacher knows different ways of explaining the same thing, knows different ways of teaching, not always standing in front of the class and talking. So, I feel like a teacher is supposed to be there to help you understand whatever subject that they are teaching to the level that they want you to understand it. In in the same sense, they teach you that knowledge that they want you to understand.

Another study participant also highlighted a specific teacher as an example of caring instruction. Specifically, Carmen, a sociology major and former biology major, said:

> if you want to talk about a caring instructor, an all-around caring instructor, I had one; Mr. Collins. He took a serious interest in me as a math student. I was feeling like a failure in some areas, but he took the

time to exaggerate the areas that I was good at. For example, to take my reading skills and bring it into math. He did not have to because that was not his field of study. Nor was that his section to teach, but he cared.

Other African American women reflected on their girlhood experiences of caring instruction and culturally responsive instruction as shown through hands-on learning and teachers caring for them as individuals. Crystal, a health science major and former biology major, shared her perspective as follows:

> For science courses, the teachers' approach was more we are going to teach you how to do this. Then, you would do it in class. You had homework. We actually got to work it out. It was hands-on. You did not have to sit there and take notes. You were writing when the teachers were writing.

Shannon, a biology major, emphasized the care that she felt from her sixth grade teacher. She explained:

> She was a math instructor. I think that I appreciated her a lot, not just because she was Caucasian, but because she really believed in all of us. I kind of experienced some aversive circumstances in high school. So, I think that I appreciate her a lot. As far as being a caring instructor, I do not know, if I can really pinpoint it. It is something that you just feel, like the way that they teach, the time that they take to teach, if they will take your questions. I know that some teachers will get frustrated if you start asking questions and you start questioning whereas caring teachers will take the time to explain it and they will not get frustrated.

Finally, a business major, Patricia described her experience as follows:

> How I learn best is with hands-on learning and activities. So, pretty much if I am engaged in the materials, then I learn best. They were sweet older ladies. Again, they were able to explain it and break it down where we could understand it on our level. They had different activities and projects. Although I did not like speaking in class, we had to do some projects and present some stuff in front of the class. It was fun stuff to do. It was actually hands-on.

These passages support the idea that caring instructors who employ culturally responsive teaching support African American girls' efforts to grasp the content of the science and math courses. They are also more hands-on instructors who cared about their students' learning.

> The African American women also described how they benefited from caring instructors who practiced culturally responsive teaching within their high school math and science classrooms. For example, Regina, a sociology major and former pre-business major, recalled, "My science and math teachers were patient. They took their time and were able to explain things clearly. I guess probably in comparison to college, it seemed like they actually cared. I mean that they took a concern in me."

A biology major, Shannon reported similar experiences in high school with both her math and science teachers. She pointed out:

> For high school math, the teachers took an interest in the students. They cared if you learned the materials. I felt that they felt passionate about it. They took their

jobs seriously. For science, they also took an interest in the students and took the time with the materials. They liked what they were doing. That is what I felt. If you needed time after school, they would stay, or they would arrange a time for you to come meet with them. Pretty much any of my instructors would do this. If you needed help or you needed some time after class, they would arrange certain days when they would tutor or meet with students if they had questions.

In a similar fashion, Lara, a math major, described the critical importance of the teaching methods of the math and science teachers. She stated:

Ability to teach or explain things so that students who are just learning the materials can understand it; that was my judgment on being a good teacher. By high school we had to know the stuff for exams. If you could teach it and really teach it in a way that I not knowing it was able to understand it and do it, then I felt like that was good teaching. That goes for both inside and outside of high school.

Further, Crystal, a health science major and former biology major described how her caring and culturally responsive science teachers in high school helped her to learn the course materials.

For science, the fact that the teachers made sure that students understood the materials showed that they cared. They did not just smack the materials on the board. Everybody had their day when they did not understand something. Everybody had their day when the instructors rushed through something. For the most part, it was like you get this, "Are you sure that you get this? Maybe we need to take another day on this tomorrow." Instead of "if you don't get it, then you need to go home and work it out yourself."

Finally, Carmen, a sociology major and former biology major described the caring instruction and culturally responsive teaching that she received from science and math instructors.

> For math, my teacher was a knowledgeable instructor. She didn't mind helping me. It seemed like she wanted me to do okay. I had the kind of teachers who, because it was with the same students all of the time, we built a bond whether they liked it or not. If we asked questions, they were going to answer the questions to the best of their ability. So, it wasn't necessarily the teaching, the way that they taught, but the way that they reviewed. My science instructors were as good as they could be. I think the smaller the class setting, the easier it is to handle, the easier it is to address every student's needs. The more that you get up into numbers, the harder it is to be able to connect with your students and get everybody's questions answered.

Moreover, the teachers of the African American girls were also responsive to their learning needs in science and math classes. For example, they used multiple teaching styles (e.g., visual, auditory, kinesthetic) to match the students' learning styles. Engineering major, Simone stated that, in general, her science teachers "had a lot of hands-on work and different ways for students to learn the materials." Kayla, a chemistry major and former health science major, described a similar experience. She stated:

> I know that for science they did a lot of in-class experiments; not blowing up anything, but the simple things like for density and gravity. They made it visual, and

that is what helped me a lot. For math, they did a lot of examples in class. At home, I did not get a lot of help with my homework so going by those examples before we got to the work really helped as well.

In middle school, Amber, a health science major and former biology major reflected positively on the instruction of one teacher. She explained:

> We kind of felt like the teacher was our friend. And we could learn, ask questions. We took a lot of notes in that class, I remember. If they wrote on the board, we took notes in our notebook. Or we did a worksheet, or we watched a movie, or we did something hands-on outside. Or, we thought outside the box. The teacher used all of the learning styles, instead of just focusing on one like in college. I'm going to lecture you for hours and hours. I'd fall asleep. I'm just saying. Like, you know, not being in that monotone voice, what the heck? I don't understand how the university thinks that's okay. Monotone, really?

In addition, Briana, a biology major, explained the effective approach of the teacher she had for both geometry and AP calculus. She stated:

> She was very thorough and detailed and broke it down so you could get the concept. She would walk you through step-by-step; this is what you do to write a proof for geometry. These are the things that you do and might want to think about when you are taking a derivative in calculus. You do it like this, or this is a shortcut you can use. This is how you remember this concept. She made things very easy to remember.

Furthermore, Raven, a health science major and former biology major, appreciated that her teachers presented material in a "way that students could understand it, like relating it to things that we see every day. That sticks in your mind more than something that you have no idea about."

Finally, African American women spoke about how teachers made their classes fun and interactive through their teaching practices, including being personable and modulating their voices. For instance, Ashley, a biology major, described two of her high school science teachers as follows:

> My chemistry teacher was a lot of fun. He always did school demonstrations in class. He was able to make chemistry easy for a lot of people who may have struggled from time to time. The teacher who taught aquatics, he was very easygoing. He was from California. He did not care about much, but you could tell that he was really interested in the material that he was teaching us. So, he had a lot of fun teaching us and we had a lot of fun learning.

The above narratives illustrate the practices of caring science and math instructors who taught the study participants science and math classes in elementary school through high school. Specifically, the caring science and math instructors were very interactive and hands-on with African American girls in math and science classes. These stories are also consistent with the literature on interpersonal caring instruction (Siddle Walker, 2001) and culturally responsive teaching in which the learning styles of African American girls are matched within science and math classrooms (Bondy et al., 2013; Gay, 2018; Howard, 2001).

College Preparation

Culturally responsive teaching and caring instruction in high school helped to prepare the African American women in this study for math at Town University. Health science majors and former biology majors, Danielle and Crystal share their lived experiences on math preparation for college. Danielle stated "I only had to take statistics, so I feel like I was overprepared for that." Similarly, Crystal explained her high school preparation for college math at Town University. When asked if her high school classes prepared her for college, she replied:

> For math, they did. I have only taken two math classes. I am in one now. One was freshman year. Other than that, I have not touched math. I have not had to touch Calculus or Calc I or anything. I have not taken any math in three years and I am doing fine. I have an A in the statistics class that I have now. It may be that, even if I don't know how to do something in math, then I am going to go home and sit there and look at that paper until I can figure out what number goes where.

Likewise, math major Jennifer said, "I definitely felt prepared when it came to math." Biology major Briana also reflected positively on her high school math preparation. She pointed out: "I had a phenomenal high school math teacher who taught me geometry and calculus. I did not put hardly any work, any effort into my college calculus class. I saved all of my notes, and I was able to coast through my calculus class during freshman year. I got a B+." Ashley, another biology major, explained why she felt prepared for college-level math classes as follows:

> The math program in high school was integrative math where instead of being given formulas to use, we had to figure out the formulas ourselves based on the work that we were doing and then apply them. So, for example, we had to derive the Pythagorean theorem from

what we were doing in class, so we understood where it was coming from and why it worked.

Although the African American women felt prepared for college math, they felt unprepared for the rigors of science at Town University. When asked whether her high school experience in science had prepared her adequately for college courses, biology major Briana replied:

> Yes and no. Yes, in that I had at least seen some of the stuff, but no, because the amount of detail that you go into in college-level science courses does not compare to what you do in high school. I took biology during freshman year in high school, and I took an introductory biology class here during my freshman year at the university. They were nothing alike. They were on two totally different levels. I had at least seen the basics. I at least had heard DNA replications, transcriptions, and translations. I knew the basics, but not in detail.

Shannon, also a biology major, similarly felt unprepared for science classes at Town University. In part, she attributed that to advice she received from an academic counselor in high school. She explained:

> I was signed up for AP Biology during my senior year, but the counselor told me that, if you are a science major, colleges do not recommend that you use your AP credit, because when you use your AP credit, you do not have to take the class in college. They know the college curriculum does not even compare to high school. So, you need to take the class once you get here. Since they said that, I did not take the AP class, but it would have helped me so much had I taken it. So, I got a little bit of background in chemistry, and in

biology, and in physics. It really is no comparison to what you really need to know once you get to college.

Another biology major, Rachelle also felt unprepared for the college science curriculum. When asked if she was prepared for the science courses at Town University, she stated: "Science, no. I got the good grades, but I did not retain any of the information. So, maybe if I had been focused on retaining, then I would have."

Crystal and Amber, both health science majors and former biology majors found science to be difficult at Town University even though they took advanced science curriculum in high school. For example, Crystal said:

> For science, no. I did not feel prepared. That was only because like I said chemistry was a big blur to me. I refused to take it over in high school, because I had a bad experience with it…Other than that, no it did not prepare me for college-level science courses.

Amber added to the conversation on the lack of high school preparation for science curriculum in college. She stated:

> I think AP biology prepared me the best that it could. I really believe AP biology at my second high school did make a difference. Although, I got a C there and I got a C in Molecules and Cells here at Town University, but I probably could have gotten a B if I hadn't had other issues and stressors, and if I had taken it freshman year when it was fresh in my mind instead of taking it sophomore year when I did. Because AP biology was so close to Molecules and Cells, my taking it during my freshman year would have given me a leverage because I had just seen all that information.

Further, despite taking advanced science courses in high school and other colleges, sociology major Carmen also felt unprepared for her science classes at Town University. She recollected:

> For instance, every college course has an introduction to that course. High school science probably prepared me for that introductory course. Of course, you automatically matriculate to the level that you feel that you are supposed to. I don't know if I felt unprepared because we did not go as into depth or if we didn't review or if I just didn't remember but I found myself in class relearning everything. I even felt unprepared after taking science courses at Brownstone University (pseudonym), a Historically Black College and University (HBCU) and transferring to Town University.

These stories demonstrate that the African American women in this study were not prepared for science at Town University. These narratives are consistent with prior research that found that students of color come to college unprepared for the rigors of the science, technology, engineering, and math (STEM) curriculum, even if they took AP or honors classes in high school (Seymour, Hunter, & Weston, 2019; Thiry, 2019).

Conclusion

African American girls developed strong relationships with their peers and received culturally responsive teaching and caring instruction within their math and science classes in K-12 schooling. The caring science and math instructors used a culturally responsive teaching style to match these students' learning styles (e.g., visual, auditory, kinesthetic). This teaching style enabled the study participants to learn and engage in math and science in primary and secondary schools. They also had access to AP and honors science and math curricula to prepare for college-level studies. While they were prepared for the math courses at Town University, they still felt unprepared for their rigorous college science courses. This lack of preparation was not due to the caring instruction, culturally responsive teaching, or enriched curriculum that they received in high school. Rather, the African American women who pursued STEM majors at Town University encountered a hidden curriculum. The concept of the hidden curriculum is discussed in the next chapter.

References

Bondy, E., Ross, D.D., Hambacher, E. & Acosta, M. (2013). Becoming warm demanders: Perspectives and practices of first year teachers. *Urban Education, 48*(3), 420-450. https://doi.org/10.1177/0042085912456846

Gay, G. (2018). *Culturally responsive teaching: Theory, research, and practice.* New York: Teachers College Press.

Howard, T. C. (2001). Telling their side of the story: African-American students' perceptions of culturally relevant teaching. *Urban Review, 33*(2), 131-149. https://doi.org/10.1023/A:1010393224120

Seymour, E., Hunter, A., & Weston, T.J. (2019). Why are we still talking about leaving In E. Seymour & A. Hunter (Eds.), *Talking about leaving revisited: Persistence, relocation, and loss in undergraduate education* (pp. 1-53). Switzerland: Springer.

Siddle Walker, V. (2001). African American teaching in the south: 1940-1960. *American Educational Research Journal, 38*(4),751-779. https://www.jstor.org/stable/3202502

Thiry, H. (2019). Issues with high school preparation and transition to college. In E. Seymour & A. Hunter (Eds.), *Talking about leaving revisited: Persistence, relocation, and loss in undergraduate education* (pp. 137-147). Switzerland: Springer.

Chapter 4

The Hidden Curriculum Within the Culture of Science in Higher Education

Chapter 4 addresses the question, "What is the hidden curriculum within STEM majors at Town University?" When transitioning from high school to college, some African American women who pursue science, technology, engineering, and math (STEM) majors struggle in the associated courses because they are unaware of an important tool required to be successful, namely, the hidden curriculum within the culture of science at predominantly White institutions (PWIs). The hidden curriculum consists of rules, norms, behaviors, and values that are part of a specific culture (Bourdieu, 1984; Feinberg & Soltis, 2009).

Oftentimes, African American women lack access to the hidden curriculum in STEM fields. This chapter describes the often unacknowledged hidden curriculum within the culture of science that discourages African American women from pursuing STEM majors at PWIs. The chapter begins by describing the tenets of the hidden curriculum in STEM majors. The hidden curriculum within STEM majors in postsecondary settings involves (1) knowing about the academic rigor and competitiveness of courses, (2) experiencing course failure and information overload, (3) learning course materials during office hours, (4) using campus resources, and (5) knowing about the departmental culture. The chapter concludes with a summary and a set of recommendations for administrators and faculty members.

Academic Rigor and Competitiveness

As part of the hidden curriculum in STEM majors, African American women encounter challenges due to the academic rigor and competitiveness within the STEM culture. The STEM majors in this study shared their experiences with the academic rigor and competitive culture found in weed-out courses at Town University. Weston, Seymour, Koch, & Drake (2019) defined gateway or weed-out courses as "a subset of introductory, foundational courses that students normally take in the first or second year of college" (p. 197). Completion of these courses is difficult "because grading is much more severe than in other university courses. Instructors may grade on a curve or give D and F grades to a predetermined quota of students" (Weston et al., 2019, p. 197).

In terms of weed-out classes, Briana, a biology major, described her experience as follows:

> In my chemistry courses, I did not perform as well on exams. In terms of knowledge, I knew as much as all of the rest of my peers. It is just that I am a bad test taker. The exams didn't show that I had the knowledge. The professors that I had were quite challenging and tough, the way that they wrote their exams. The General Chemistry classes here at the university are weed-out courses. They are designed to be overly difficult to weed people out. In my biology courses, I have done very well actually. I have gotten B's in all of my biology courses.

When asked to identify a weed-out class she had taken at Town University, Shannon, also a biology major, said:

> It was precalculus. I got a C. When I took calculus at a junior college, I got an A. I think that precalculus

does not have a lot to do with calculus, like you would think. I don't know. I did not really like precalculus. It might have been the pace of the class. I had already taken calculus in high school.

Shannon noted that there was a curve and grading scale in the precalculus class as well. Another biology major, Rachelle, asserted that:

> Molecules and Cells and Genetics are definitely weed-out classes. If you don't study, you're not gonna make it. I don't even feel like the material is that hard. It's just, if you don't study you won't make it. It's a big weed-out. From biology, some of my friends went to English. I think one is still a biology major with me. That's because they don't have the drive. They don't want to do it. It takes a lot of work, a lot of studying. And they just don't want to do it. So, they're not biology majors anymore.

These statements indicate that the African American women in this study were not put in a position to fully learn the STEM course materials because of the competitive structure of the academically rigorous courses that were set up to weed-out, and thus fail some students while keeping other students in science majors. This finding is similar to research by STEM scholars (e.g., Seymour, Hunter, & Weston, 2019). Seymour et al. (2019) posited that:

> the absence of apparent structure in the selection of class materials, the order, and logic of their presentation; and lack of fit between class materials and homework, and the content of tests, all suggested that instructors knew little about organizing their teaching and learning objectives that were shared with their students. (p. 11)

The competitive culture also forced the African American women to compete with their peers to earn good grades through a curve. Seymour et al. (2019) further noted that "curve grading was portrayed as the engine driving competition because it forces students to compete with each other by exaggerating fine degrees of difference in performance" (p. 15).

Course Failure and Information Overload

The African American women also found that the hidden curriculum within the STEM culture at Town University made them prone to failing science courses because of a survival-of-the-fittest attitude that relies on information overload. For example, Carmen, a former biology major who switched to sociology, described her experience as follows:

> As soon as I got to Town University, it was a different atmosphere, more pressure. Like, I said, I do not like pressure. Not that I crumble under it. I still maintain. If you were to look back at my transcript from my first year, you would have A's sitting next to F's; well, now they are W's [withdrawals]. That was a hard year. Come to find out that for the first Molecules and Cells class, there are more students that fail that class than pass it.

Similarly, Danielle, another former biology major who switched to the health science major discussed information overload. She stated "Introduction to Physics was kind of rough, and then General Chemistry I and General Chemistry I Laboratory just come at you so fast with information. I guess coming from who I am, I'm just one to push through." These narratives point to course failure and information overload within the STEM culture at Town University. This finding is consistent with prior research on learning in STEM majors according to which female students are likely to switch majors due to fear of future failure (Holland, 2019). Moreover, Weston et al. (2019) pointed out that

> among the testing and grading practices described as key characteristics of weed-out course experiences are lack of alignment between testing questions and emphases in class content and misaligned levels of difficulty between homework and exams…Assessments also trip students up where they do not address core concepts. (p. 202)

This practice leads students to be in a survival-of-the-fittest mode, which means that some will leave STEM majors whereas others will remain in these majors (Weston et al., 2019).

Office Hours

To survive in STEM majors, the African American women in this study learned about an aspect of the hidden curriculum at Town University that involved attending office hours to learn the math and science course materials. During office hours is when they also built relationships with their instructors in a smaller setting. For example, Carmen, a sociology major and former biology major, described making frequent use of office hours. She stated that: "before I came to Town University, I was really adamant about attending office hours. I used to always go to my student/teacher office hours, my teaching assistant's office hours, and so forth." She continued to say that:

> I keep saying this, I have never had an issue with science until I got here to Town University. I did pretty well. I still talk to my previous teachers. At an Historically Black College and University (HBCU), Brownstone University (pseudonym), I tutored students in biology and chemistry. So, I was cool with most of the students because I saw them in the lab. Here at Town University, I started off cool with the people that I met in my chemistry courses, but then we fell out because they had bad attitudes.

Briana, a biology major, also attended office hours to learn the materials for science classes. She pointed out:

> For science classes, I ask questions in office hours. I feel very comfortable going up to a professor if he seems very open, very friendly, willing to help you, like an ask-questions-if-you-need-help type of person. If he is that type of professor, sure I will go to office hours the entire time and say, 'Hey, I did not understand this.'

Shannon, another biology major, further stated her reasons for attending office hours. She said: "A lot of my friends, if they have a burning question or something, will just hold it until they go to office hours for science classes. I don't like to interrupt, so they might have the same reasoning."

In a similar fashion, the African American women attended office hours in the math department. For example, biology major Briana replied:

> The two times that I went to my Calculus I teacher's office hours, she was very helpful, very nice. She broke everything down and gave us sample problems.

She made me do a problem in front of her to make sure I got the concept. I liked her a lot. I really did, but I just didn't really care for math. I was just doing it because I had to. My Calculus I teacher was a middle-aged, White female. She is very nice, very sweet, very friendly, a very good teacher. She is very good at explaining things to you in detail.

Engineering major Simone also reported attending office hours for her math classes. She said: "I saw the instructors in the class or in their office hours. And that's the main way I talked to them." In addition, math major Lara found office hours to be important in mastering the concepts taught in her math classes. She explained:

> For math problems, I felt comfortable asking questions, because I didn't like not knowing something, so I would go to the professor's office hours pretty much any time I could make them, definitely every week, to get help with the extra stuff that I didn't completely understand the first time.

According to these stories, attendance at office hours is a critical part of the hidden curriculum at Town University. Office hours are when African American women in this study learned the course materials and asked questions to fully understand the information presented in their math and science classes.

Campus Resources

To remain in STEM majors at Town University, the African American women in this study learned about another aspect of the hidden curriculum in the STEM culture. This included using campus resources, such as studying with peers who did not look like them and were not their friends to succeed in STEM majors. Several of the study participants provided recommendations for resources future African American women students to pursue while studying in STEM fields at institutions like Town University. Engineering major Simone, for example, shared the following general advice: "I think that they are going to have to work harder, talk to teachers, make friends with people who did take those classes in high school, and just work their butts off."

More specifically, biology major Briana listed several resources with which she had personal experience. She pointed out:

> Your professors are a good resource. I feel like a lot of professors are more sensitive and more understanding if you talk to them one-on-one. Try the Counseling Center, if you just need a third person to vent to. The Minority Resource Center, because they tutor in a lot of general education courses. For biology help, there's a tutoring group. Attend office hours as much as you can for all of your Molecules and Cells courses. There are also a ton of churches and church-associated fellowship groups on campus.

Shannon, another biology major, also encouraged future African American women in STEM majors to explore campus resources, including other students. She responded by saying that:

> This school has pretty much unlimited resources. If there is ever a thing that a student feels could be useful to them, then they should look it up. Try and see what is out there. Definitely get a tutor, because regardless whether you are getting an A or a D, they are going to help you. These are students who already went through

these classes, so they already know tricks, little things to help you learn information better...I think mentorship is a good thing.

Support for developing relationships with fellow students was reinforced by biology major Ashley, who pointed out that:

> Hopefully, at some point, there will be more mentors who look like them. At this point, you just have to make connections with who you can. Find someone who you look up to who you think knows what they are doing and what they are talking about, you enjoy what and how they are teaching you, and you feel that you can really learn from that person. No matter what they look like, you should do this because if that person is where you are or where you are trying to be right now, then there is no better person to learn from.

These recommendations reveal that African American women sought out campus resources (e.g., professors, teaching assistants, tutors) to learn course materials in their respective STEM majors. Additionally, the comments about mentors are consistent with findings in the literature on the lack of African American women in STEM career fields (Tickles and McPherson, 2016). As a result, in STEM majors and careers, African American women have fewer opportunities to be mentored or to have role models of their own race and gender (Borum & Walker, 2011; Brown, 2011; Dortch & Patel, 2017; Malcom & Malcom, 2011; Tickles & McPherson, 2016). Accordingly, they may have to navigate the culture of science in isolation and may be unaware of the hidden rules that they need to know to successfully navigate the culture of science at the undergraduate level (Bryant, 2019; Ong et al., 2011).

Departmental Culture

The African American women in this study learned the hidden curriculum of their respective STEM departmental cultures. For those in STEM majors, these cultures were characterized by (1) isolation, (2) a chilly climate, (3) minimal opportunities to engage in research projects, (4) psychological stress, (5) the need for self-teaching, (6) a mismatch with student learning styles, and (7) a lack of faculty diversity. Each of these characteristics is addressed in turn in the remainder of this chapter.

Isolation

The African American women in this study reported feeling isolated because of the demographic backgrounds of their peers in their science and math classes at Town University. For example, a former biology major and now a health science major, Crystal discussed the shortage of African American women in science classes. She said, "It's probably 10/90 compared with female students with 10 being African American women." Another former biology major and now a health science major, Danielle mentioned being one of the few Blacks in a biology class. She stated, "Biology was a mixed–gender class. I was the only Black person in there." Likewise, Ashley, another biology major, expressed a sense of isolation in her science classes when she said: "sometimes, I am the only one. I am the only Black person; people will doubt my answers."

In addition, a biology major, Shannon reported isolation in the science courses at Town University, when she responded:

> How many of my peers looked like me? I will do percentage wise. I will say like 5%. Well, these classes are

huge. Black women, I'll say a Biology lecture is five hundred fifty people, there will maybe like 10 Black women. My best friend and I have the same major.

Similarly, a biology major, Briana reported her experiences in science classes. She stated:

In my general chemistry or biology courses, probably no more than 10 students are Black. They have been weeded-out. In my particular advanced biology courses, there are four. It is discouraging because there is nobody in my classes who look like me. I mean there are very few. It is just difficult not being able to be with your friends. So, they can't help you.

The African American women in this study also experienced feelings of isolation in their math courses. For instance, math major Jennifer commented, "Yeah, the number of other Black women in my classes is usually zero. There are no Black girls who want to be math majors." Shannon, a biology major, contrasted her experience at a community college with that at Town University. She explained:

For math, at the community college, there were about two to three White men. The rest were African American men and women. At Town University, I noticed a few more African Americans than Whites at community college. I would estimate that 10% rather than 5% are African Americans, and for African American women, it might be 5% of about 150 people.

All of these responses point to the lack of a sense of belonging and feelings of isolation, which are common experiences among African American women pursuing science majors (Bryant, 2019; Brown, 2011; Dortch & Patel, 2017; Ireland et al., 2018; Jordan, 2006; Moore, 2016).

Chilly Climate

In addition to the isolation that the African American women in this study endured at Town University, they also encountered chilly cultural attitudes of people within their departments. For example, biology major Shannon shared her impression of the campus culture. She stated "they discourage you. They try to tell you, "You are going to change your major." I had people tell me that before I even got here." Shannon also recalled an experience in her department that reflected a discouraging, rather than supportive, attitude from a fellow student. She pointed out:

> Last semester, I had a lab partner who I did not care for. She was White and a transfer student. She had a miss know-it-all attitude. That was my first biology lab, so I was not as familiar with the techniques and stuff. I was not the only student in their first biology lab. I guess at her school, it was pretty much standard for students to have already taken a biology lab, so she got a little frustrated with me.

Another biology major, Rachelle, also shared a negative interaction with other students in a laboratory class. She said:

> In lab, some girls didn't want to help with the lab when I asked the group next to me. I believe they were White. They were the really, really nerdy girls with long ponytails that goes to their tush with the glasses and the pants pulled up to the waist. I think that's the only time I've ever had an issue with someone not answering my question.

Similarly, engineering major, Simone described a chilly peer culture within her department. She pointed out:

> Definitely had to invite myself to be included in groups. It was a little bit more difficult to find groups to work with in classes, so I had to make sure I was there on the day when groups were formed.

These accounts suggest that chilly climates made African American women feel discouraged and excluded within their respective STEM majors, especially when doing group work.

Research Projects

African American women in this study reported fewer opportunities to engage in research projects as biology and chemistry majors at Town University. For instance, Amber, a former biology major, and a current health science major, also spoke about the lack of exposure to research opportunities at Town University. She remarked, "I have always wanted to do research. Actually, I talked to a couple of professors about research and the research that they do, but by the time that they started it, I didn't have the time to do it." As a result of this lack of access to research opportunities with professors, African American women might have to seek opportunities outside of their departments or colleges to obtain research experience.

Psychological Stress

African American women also dealt with psychological stress when pursuing STEM majors. For example, Danielle, a health science major and former biology major, acknowledged, "Sometimes, it was stressful. It was a lot of work. There was a lot of reading that I did not do. Looking back, I made it through. I am happy that it is over." In contrast to this story of successfully overcoming elevated stress, the experience of Carmen, a sociology major and former biology major, provides a concerning illustration of the negative effects such stress can cause. She replied:

> In my first nine months here, I literally went from 170 pounds to 115 pounds. I had lost so much weight. I would cry so hard that I would vomit. It was bad because I had never done that poorly in school. I got two D's the second time I took two classes. It was ridiculous. So, Town University dropped me, and my scholarship dropped me. Now, I pray or I cry myself to sleep. I have been known to spas out until everything goes black. My hair fell out…When I feel isolated, I sleep. When you sleep, you don't know that you are alone. When I am sad, because of loneliness or whatever, I shop, which is definitely not good because I can't afford it.

These reports are consistent with the STEM education research on stress management during the pursuit of an undergraduate degree (Morton & Parsons, 2018).

Self-Teaching

Within the hidden curriculum of STEM at Town University, the African American women in this study found that they had to teach themselves science concepts because they were not covered thoroughly in class. For example, biology major Briana described how the culture of science resulted in her having to learn chemistry concepts without assistance from her instructor. She explained:

> In the Department of Chemistry at Town University, the professors are like either you sink or you swim. For Organic Chemistry, my professor prerecords all of the lectures. All of the lectures are online, and students are responsible for going online three times a week and watching the lectures, which are at most 10 minutes long. That does not substitute for the teacher being in front of a class, teaching me, explaining to me what is going on, and giving me feedback. I feel like my instructor is very cold and very distant. If you want to see him, you make an appointment. He does not have office hours. In this class, you have to use a computer-based program to submit all of your answers for every quiz and exam and then just get discouraged because your answer is incorrect, so you just move on and try again.

Rachelle, another biology major, also described having to teach herself the materials in a science course at Town University. She noted,

> My Physics professor here at Town University did not teach that material. Not just me, but everyone who took the class was not happy with the material that we were given. The physics professor at Town Community College was amazing. The work wasn't any easier than it would be here. It was the same difficulty level, but the way he taught it to the class really helped.

These narratives highlight the poor teaching styles encountered within STEM classrooms at Town University that did not support learning by African American women in science and math classes. Similarly, Seymour et al. (2019) found that "students were frustrated by instructors who seemed unable to explain their material sequentially, coherently, or break it down into sequences that would enable conceptual grasp" (p. 11). This type of instruction also points to instructors' lack of caring for students' learning within the classroom (Seymour et al., 2019).

Students' Learning Styles

Several study participants reported failing to learn the science course materials because of the departmental teaching styles within the culture of science at Town University. When asked whether the teaching style in science matched their learning styles, some of the African American women in this study identified a poor match between the teaching styles they experienced in science classes and their learning styles. For instance, Briana, a biology major, explained that "understanding a class lecture is more of a responsibility on the student, because the professor will keep going." She continued to say that:

> I have had professors announce in class, "I will keep going if you don't stop me to ask questions." If you don't, you will fall behind. Basically, going to office hours, hands-on interaction with the professor, and hands-on interaction with the concepts that we are reviewing are how I learn best. I have had several professors, for the most part all of the professors that I have had for biology and chemistry, who you have to talk to one-on-one. They will break out whatever models they have to make you understand it a bit better. That really is what works for me to better learn the concepts is that hands-on interaction in office hours that helps me learn or see whatever concept that I need to see in person.

Shannon, another biology major, also felt that the approach used in science lectures, as well as that used in laboratory classes, did not match her learning style. She stated:

> It is pretty much you go to lecture and you get the material. If you want to grasp the concepts and have questions and stuff, the appropriate setting for that stuff is outside of the classroom in office hours or in your discussions, meeting up with your TA's [teaching assistant's], or something like that. Lecture, I really don't think that you learn in lecture. For the most part, the labs have not been matching up with what we have been learning in lecture. Once students are done with the lab, they turn in the lab, and they still don't understand what they did. So, I would say no, the teaching style does not match the students' learning styles.

A similar viewpoint was offered by Carmen, a sociology major and former biology major. She explained

> I am a hands-on learner, so for my learning style, no the teaching style is not a good match. I need to be able to see the problem, step-by-step. That is how I need to be able to see it, so I can mimic the steps. In biology, you cannot do that because it is pretty much reading and memorization. So, I just have to read things over and over.

These responses show that the African American women STEM learners in this study took on the responsibility of learning science by going to professors' office hours to fully learn the materials. The lecture style setup of the science classes did not match their learning styles, so they had to hold themselves accountable for learning to continue pursuing STEM majors at Town University. As reported by Seymour et al. (2019), African American women students in STEM majors look for hands-on learning and real-life applications to be successful in science classes. However, the setup of the lecture and laboratory courses at Town University did not provide African American women with the opportunity to fully grasp the materials in class because of a mismatch with their learning styles.

Faculty Demographics

The African American women in this study experienced departmental cultures at Town University that lacked faculty diversity in math and science. For example, Lara, a math major, summarized her math teachers as follows: "I had a White man, an Asian man, another White man, a White woman, a Turkish man, and a Russian man, so mostly White men." Simone, an engineering major, reported, "for math, I had mostly White, female professors, a couple White men, and one Indian guy."

The faculty makeup in the science departments at Town University also lacked diversity. For example, biology major Briana said:

> In chemistry, I had all White men, older, middle-aged White men. In biology, I had White men, middle-aged. I had two middle-aged Asian men. I have also had middle-aged Indian men. I did take physics, but I took it at Town Community College (pseudonym) with a White male instructor in a class with 30 students.

Shannon, another biology major, added "I had one female chemistry professor during my first semester here. For biology, my professors have all been White men; I have had no female instructors in biology."

These demographics indicate that the African American women at Town University experienced learning science and math in environments without instructors who looked like them. This suggests a need to diversify the faculty members in STEM environments at PWIs.

Conclusion

African American women in this study who persisted in STEM majors at Town University learned to navigate the hidden curriculum within the STEM culture. This hidden curriculum revolved around (1) the academic rigor of the courses and competitiveness within the STEM culture, (2) the widespread experience of course failure and information overload, (3) the practice of learning course materials in office hours, (4) the need to make use of campus resources, and (5) the chilliness of departmental culture. The departmental climate was one in which some African American women faced isolation from peers, endured psychological stress, and lacked access to research opportunities. They also experienced learning environments that failed to match their learning styles, so they had to teach themselves the course materials. Finally, they were in STEM departments that lacked racial diversity as well.

The hidden curriculum created racial and gender inequalities by failing to provide equal opportunities for African American women to learn in STEM at Town University. As a result, a recommendation is to make the hidden curriculum visible to African American women pursuing STEM majors at PWIs like Town University. Additionally, postsecondary settings, especially PWIs, should consider diversifying their faculty members to hire more underrepresented scholars (e.g., African American, Latino/a, Indigenous Peoples) who understand the needs of African American women and women of color who are navigating STEM fields. Such faculty members can be recruited through

STEM doctoral programs and industry. They can serve as professors, mentors, and role models who can assist women of color and promote their college persistence in STEM fields. Moreover, postsecondary institutions might want to consider providing African American women with access to faculty members who can serve as mentors and help them become involved in faculty research as well (Ashford et al., 2017; Chang et al., 2014; Hurtado et al., 2011, Tickles & McPherson, 2016).

References

Ashford, S. N., Wilson, J. A., King, N.S., & Nyachae, T. M. (2017). STEM SISTA spaces creating counterspaces for Black girls and women. In T. S. Ransaw and R. Majors (Eds.), *Emerging issues and trends in education*. Lansing: Michigan State University Press.

Bourdieu, P. (1984). *Distinction: A social critique of the judgment of taste*. Cambridge: Harvard University Press.

Borum, V., & Walker, E. (2012). What makes the difference? Black women's undergraduate and graduate experiences in mathematics. *The Journal of Negro Education, 81*(4), 366- 378. doi:10.7709/jnegroeducation.81.4.0366

Brown, J. (2011). *African American women chemists*. New York: Oxford University Press.

Bryant, T. (2019). *Unhidden and unrelenting figures: The persistence of Black women in STEM disciplines* [Doctoral dissertation, California State University, Long Beach]. ProQuest Dissertations and Theses.

Chang, M. J., Sharkness, J., Hurtado, S., & Newman. C. B. (2014). What matters in college for retaining aspiring scientists and engineers from underrepresented racial group. *Journal of Research in Science Teaching*, *51*(5), 555-580. https://doi.org/10.1002/tea.21146

Dortch, D. & Patel. C. (2017). Black undergraduate women and their sense of belonging in stem at predominantly white institutions. *NASPA Journal About Women in Higher Education*, *10*(2), 202-215. https://doi.org/10.1080/19407882.2017.1331854

Hurtado, S., Eagan, M. K., Tran, M. C., Newman, C. B., Chang, M. J., & Velasco, P. (2011). We do science here: Underrepresented students' interactions with faculty in different college contexts. *Journal of Social Issues*, *67(3)*, 553–579. doi: 10.1111/j.1540-4560.2011.01714.x

Ireland, D. T., Freeman, K.E., Winston-Proctor, C.E., DeLaine, K. D., Lowe, S. M., & Woodson, K. M. (2018). (Un)hidden figures: A synthesis of research examining the intersectional experiences of Black women and girls in stem education. *Review of Research in Education*, *42*, 226–254. https://doi.org/10.3102/0091732X18759072

Feinberg, W. & Soltis (2009). *School and society*. New York: Teachers College Press

Jordan, D. (2006). *Sisters in science: Conversations with Black women scientists on race, gender, and their passion for science*. West Lafayette, IN: Purdue University.

Malcom, L., & Malcom, S. (2011). The double bind: The next generation. *Harvard Educational Review*, *81*(2), 162-172. https://doi.org/10.17763/haer.81.2.a84201x508406327

Moore, J. S. (2016). *Do I belong? Narratives of sense of belonging and fit from underrepresented African*

American and Latina women in science undergraduate majors [Doctoral dissertation, Rowan University]. Rowan Digital Works.

Morton, T., & Parsons, E. (2018). #BlackGirlMagic: The identity conceptualization of Black women in undergraduate STEM education. *Science Education, 102*(6),1363-1393. https://doi.org/10.1002/sce.21477

Ong, M., Wright, C., Espinosa, L. L., & Orfield, G. (2011). Inside the double bind: A synthesis of empirical research on undergraduate and graduate women of color in science, technology, engineering, and mathematics. *Harvard Educational Review, 81*(2), 172-208. https://doi.org/10.17763/haer.81.2.t022245n7x4752v2

Seymour, E., Hunter, A., & Weston, T.J. (2019). Why are we still talking about leaving In E. Seymour & A. Hunter (Eds.), *Talking about leaving revisited: Persistence, relocation, and loss in undergraduate education* (pp. 1-53). Switzerland: Springer.

Tickles, V. & McPherson, E. (2016). Mentoring our own: African American women engineering. In K. E. Tassie & S. M. Brown Givens (Eds.), *Women of color navigating mentoring relationships: Critical examinations* (pp. 95-113). Landham, MD: Lexington Books.

Weston, T. J. Seymour, E. Koch, A. K. & Drake, B.M. (2019). Weed-out classes and their consequences. In E. Seymour & A. Hunter (Eds.), *Talking about leaving revisited: Persistence, relocation, and loss in undergraduate education* (pp. 197-243). Switzerland: Springer.

Chapter 5

Invisible in High School and College: African American Girls and Women's Support Networks

Chapter 5 considers the question, "How does social capital influence the persistence of African American girls and women in science, technology, engineering, and math (STEM)?" African American girls and women have social capital in terms of support networks in high school and college. This chapter begins with a discussion of the literature on the support networks of African American girls and women, as well as the social capital framework. The support networks among African American female teenagers in high school are described next, followed by those among African American women in STEM, social science, and health science majors in college.

African American Girls' High School Support Networks

African American girls rely on support networks to successfully navigate their way through high school. The literature demonstrates that the parents and peers of Black girls influence their college attendance and completion rates (Dixson & Chambers, 2009; Hanson, 2009; Jez, 2012). In addition, Black girls assist their peers with studying in high school (Ashford et al., 2017; Hanson, 2009). They also encourage peers to work together to obtain grades that will allow them to enter and complete college. Moreover, family members and friends support Black girls who have science interests (Hanson, 2007, 2009; Ireland et al., 2018;

McPherson, 2014; Parker, 2013). For example, parents expose Black girls to science through museum visits and math or science competitions (Fries-Britt & Holmes, 2012; Ireland et al., 2018; McPherson, 2014).

Support from teachers, peers, and parents prompts African American girls to take advanced placement (AP) and honors math and science classes in high school to prepare for STEM majors in college (Borum & Walker, 2011; Britt et al., 2010; Coneal, 2012; Fries-Britt & Holmes, 2012; Koch et al., 2019; McPherson, 2017; Young, Feille, & Young, 2017). Local and national programs, such as the Meyerhoff Scholars Program, cultivate African American girls' science and engineering interests by exposing them to an enriched STEM curriculum and encouraging them to pursue STEM degrees (Ashford et al., 2017; Britt et al., 2010; Fries-Britt & Holmes, 2012; Simmons II, 2013). Finally, parents of African American girls believe their daughters will successfully complete college (Jez, 2012) and STEM degrees (Hanson, 2009; Parker, 2013).

African American Women's College Support Networks

Undergraduate African American women with strong support networks are more likely to persist and complete college (Anderson, 2019; Banks, 2009). Having a support network also makes it easier for them to cope with academic and social challenges and finish college. For instance, African American women endure racism and sexism in interactions with faculty members and peers, particularly those who are White (Anderson, 2019; Gold, 2011; Robinson & Franklin, 2011; Winkle-Wagner, 2009). Some White faculty members and peers also expect less of Black women students because of negative stereotypes about their academic abilities. These experiences make some African American women feel that they are invisible and lack a voice. However, they continue to persist in college with assistance from social support networks, including faculty members, peers, parents, student organizations, and religious organizations (Anderson, 2019; Banks, 2009; Ceglie, 2013; Gold, 2011; Robinson & Franklin, 2011; Walpole, 2009; Winkle-Wagner, 2009).

In a similar manner, parents, grandparents, and fictive kin function as support networks for Black women in college. They hold high expectations for African American women to encourage them to remain in and complete college (Anderson, 2019; Robinson & Franklin, 2011; Williams, 2011). Religious organizations also serve as support networks for undergraduate African American women by providing religious support through prayer from pastors and congregation members (Ceglie, 2013; Morton & Parsons, 2018; Walpole, 2009). The next section considers the support networks of African American women in STEM majors.

African American Women Collegians in STEM Student Support Networks

Few recent studies have discussed the support networks that Black women college students use to successfully navigate through STEM fields. The small amount of existing scholarship on this topic, however, indicates that African American women in STEM majors benefit academically from the support of parents, peers, and student organizations (Ireland et al., 2018; Parker, 2013; Perna et al., 2009). For example, parents provide African American women with emotional, financial, and spiritual support. Religious institutions offer Black women spiritual support through pastors and church members' prayers while Black women are navigating their way through STEM majors in college (Ceglie, 2013; McPherson, 2015). Additionally, STEM peers are a part of Black women's support networks in college by providing them with opportunities to become academically and socially engaged through tutoring, study groups, and student organizations. Academic engagement and social engagement are important for college persistence and college completion (Kuh, 2009a; Tinto, 2012). Finally, some professors support Black women by meeting them during office hours to assist them with homework assignments and research projects (Fries-Britt & Holmes, 2012).

Social Capital

Social capital brings together people (e.g., community members, peers) through social networks inside and outside of organizations (Coleman, 1990; Yosso, 2005). Loury (1989) posited that social capital provides people with "nontransferable advantages [at] birth that are conveyed by parental behaviors bearing on later-life productivity" (p. 272). This statement supports the idea that a person's family and community provide resources that can eventually place a child, adolescent, or adult in an advantageous position in educational or employment settings (Coleman, 1990; Loury, 1989).

In school or the workplace, children, adolescents, and adults have opportunities to develop peer networks that function as smaller communities for obtaining information and resources that can help them become successful, advance educationally, and get promoted in the workplace. In addition, social capital provides people with an understanding of the preexisting norms and expectations that exist within families, communities, and organizations (Coleman, 1990). In urban communities, social capital plays a role when high-achieving Black youth seek out resources such as family encouragement to complete high school (Conchas, 2006) and peers to serve as role models and mentors in college (Harris, 2019). They receive academic support from peers and instructors and build relationships with peers, teachers, and counselors, through which they can obtain information and resources that promote academic success.

In postsecondary settings, the culture of science centers on teaching and learning about science through a Western framework of thinking in which faculty members are predominantly White men (Harding, 2006; Ireland et al., 2018; Lee & Luykx, 2006). Science has a distinctive culture with specific norms and behaviors (e.g., time management, rigorous coursework, group study). Becoming academically successful may be more difficult for underrepresented groups in STEM, such as people of color and women, who may lack prior exposure to the culture of science. To uncover Black women's support networks in STEM fields, this chapter employs social capital as a framework to identify the support networks in STEM fields of African American girls in high school and African American women in college.

African American Girls' Support Networks for Students Interested in STEM in High School

From a social capital perspective, some of the African American women sought out parental support to help them succeed in advanced mathematics and science classes in high school. For example, Jennifer, a math major, was heavily reliant on her parents as resources who taught her about study skills and organizational skills to help her succeed in her AP classes. She explained, "if I have a test coming up and I'm worried about running out of time, my dad might say, 'Do the easy ones first, because those will go fast, and then do your hard ones.'" Similarly, a sociology major, Carmen sought out support from her parents to be successful in math and science classes in high school. She recalled: "My parents would have given me the first lessons in study skills, even though I did not adopt those things. Now, I find myself trying to master study skills." These findings are consistent with the literature on how parents of African American girls support their daughters' high school success in math and science (Hanson, 2009; Koch et al., 2019; Parker, 2013; Young, Feille, & Young, 2017).

Other African American women in this study obtained support and information about learning how to succeed in math and science from high school instructors. For example, Patricia, a business major, stated, "our math teacher during freshman year taught us about discipline and how to think critically in math classes." Similarly, Crystal, a health science major, described her experience with a high school math instructor as follows:

> My trigonometry teacher was supportive. He was very successful. He would say, 'Do your best.' He would tell us that 'you have to work hard for what you want.' He challenged us. If he finished the lessons early, he did not let us out of class. He told us something based on his life about how he is able to do the things that he is able to do now because he had to do what he had to do to overcome challenges before.

In addition to identifying teachers as a source of support, another health science major, Danielle, named herself as a support. When asked to describe who made her successful in her high school math and science courses, Danielle responded, "teachers and myself. Knowing what works for me and what does not work for me." Studying and going through the materials, paying attention and taking notes were strategies that Danielle employed to achieve success in her classes.

High school counselors were also resources that helped advise African American women how to be successful in high school. Briana, a biology major, stated that her academic counselor "told me there is this resource. You can go to tutoring. You can go to this person. I was very close to her, the academic advisor." These accounts indicate the important role that teachers and school counselors play in supporting African American girls with STEM interests. Previous research has shown that teachers' support is crucial to supporting African American girls' success in STEM as well (Koch et al., 2019; Young, Feille, Young, 2017). School counselors also help students choose science and math curricula to prepare for STEM majors in college (Falco, 2016).

Student organizations can also provide African American girls with social networks of peers with similar math and science interests in high school and college. Amber, a health science major, and Shannon, a biology major, belonged to an organization called Math Champs. Shannon further recounted her experiences in Math Champs.

> [1]Math Champs is a group for students who are interested in math. You compete against other schools. We met once a month. We would go to competitions and compete with other schools doing math problems. It was not like a buzzer-type competition, but you would take a test, and they would see what school got the highest

---

[1] This excerpt is taken from McPherson, E. (2014). Informal in science, math, and engineering majors for African American female undergraduates. *Global Education Review*, *1* (4). 96-113.

scores. We made it through the regionals. So, we went to the state competition that was 30 minutes away. It was still in Central City.

Other African American women in the study also participated in academic and social student organizations with peers who supported their math and science interests. These experiences are consistent with prior research centered on Black female teenagers using support networks (e.g., parents, peers) to navigate through the high school curriculum (Dixson & Chambers, 2009; Hanson, 2009; Harris, 2019). They also support scholarship concerning the importance of using advanced science and math curricula to prepare African American women for STEM fields in college (Borum & Walker, 2011; Britt et al., 2010; Coneal, 2012; Fries-Britt & Holmes, 2012; Young, Young, & Ford, 2017).

African American Women's Support Networks for STEM Majors in College

This section describes the support networks used by undergraduate African American women in STEM majors at Town University. For African American women, one source of support in college from a social capital perspective was the buddy system. The buddy system was commonly utilized by Black women who persisted in math, biology, and engineering majors. The women associated with diverse peers as study buddies. More specifically, they relied on peers who were male, female, African American, White, Asian, and Indian or had other demographic backgrounds to study together in order to successfully pass upper-level courses in math and science at Town University. For example, Jennifer, a math major, recalled the buddy system that she utilized in math classes at Town University.

In a similar fashion, Kayla, a chemistry major, used her living learning community at Town University to work on a research project focused on Black women and their neighborhoods. When asked about feeling welcome in the research group, she stated, "I did indeed. All of the supervisors were really helpful. They answered all the questions thoroughly. Peers also helped out, too. They were more experienced, so I felt comfortable talking to them." These responses suggest that peers are important for providing information and resources that can promote academic success inside and outside of the classroom. They also support the idea that the high-impact educational practice of research promotes college persistence among students of color (Griffin et al, 2010; Hurtado et al., 2011; Kuh, 2009b; Moore, 2016; Perna et al., 2009).

Many African American women in this study found support through student organizations at Town University as well. Biology major Shannon described her experience in a student organization as follows:

> [2]For science, I am in an international club that supports children. We volunteer, so we don't do science activities, but we are all science majors. It is pretty much an effort to get medical attention to kids in underserved countries. We are all biology majors or pre-med. We do not really do science-related activities outside of school.

When asked about student organizations, Kayla, a chemistry major and former health science major, replied, "I know there's tutoring, as well as organizations that teach

---

[2] This excerpt is taken from McPherson, E. (2014). Informal in science, math, and engineering majors for African American female undergraduates. *Global Education Review*, *1* (4). 96-113.

you what you can do with your major." She added, "I'm in NOBCChe, which stands for National Organization for the Professional Advancement of Black Chemists and Chemical Engineers. It's geared more toward chemical engineers, but there are chemists in it too."

Organizations such as these allowed for African American women to build communities among STEM majors while simultaneously reducing feelings of isolation at Town University. Prior research has also shown the importance of student organizations in serving as support networks and promoting social and academic student engagement (Anderson, 2019; Chambers & Walpole, 2017; Kuh, 2009a; Tinto, 2012; Walpole, 2009; Winkle-Wagner, 2009).

Parents are a third type of support network for African American women pursuing STEM fields in college. For example, Rachelle spoke about her reliance on parental support from her father when she was thinking about leaving her biology major. She said: "My dad. He is really supportive." Patricia, a business major, also told of her mother, family support, and mentors. These individual experiences indicate that parental and mentor support is crucial for African American women in college, particularly during the times of greatest difficulty. African American parents provide their daughters with encouragement, hope, and the confidence needed to persevere in STEM majors. Mentors offer emotional support during challenging times as well. Prior research has demonstrated the importance of support networks in providing African American women with multiple resources that will help them persist and graduate from college as well (Anderson, 2019; Ireland et al., 2018; Robinson & Franklin, 2011; Williams, 2011).

Communities also provide collective support for African American women in STEM fields in college. When asked about sources of support for success in college, a math major, Lara responded, "my mom, counselors, graduate assistants, mentors, a lot of people collectively." When she was probed about lessons learned about pursuing a STEM major in college, she stated:

> I've learned that it is hard work and that it is not going to be easy. If you have the determination to do what you want, then you can do it. It is not going to be that easy. Networking and finding people who have similar goals to help you get where you need to be.

Lara's lived experience illustrates the important role of teachers and family members as sources of support for Black women in college. Especially significant is the role of African American mothers who give their daughters the type of nurturing, caring, and supportive environment that is missing from the relationships that African American women have with their college professors. In addition, Robinson and Franklin (2011) noted that undergraduate African American women often view mentors "as an extension of their family or community" (p. 32). Mentor support centers on mentors (1) holding high expectations of each mentee, (2) providing mentees with guidance on how to navigate through college, and (3) cheering for mentees' successes (Ashford et al., 2017; Robinson & Franklin, 2011; Williams, 2011).

Although the numbers of African American women in STEM fields in college are small, a few study participants reported acquiring social capital from professors at Town University. A math major, Lara described a mathematics professor who provided her with information on math success in college. She stated:

> My Calculus III teacher taught me ways to help get the information needed to succeed in class, such as writing notes and going to office hours. You should start doing

that in the beginning of the term, instead of waiting until you did not do as well as you wanted to have done. If you do it that way, then you have a cushion just in case you mess up.

Similarly, when asked how she learned how to become successful in her major, biology, Ashley responded:

> I think mainly my professor and grad student that I have been working with in the lab helped me. My professor has already been successful, so she is always helping me out, teaching me about writing the thesis. Collectively, all of the professors always throw out little tips.

These stories demonstrate the important role that professors play in facilitating the retention of African American women by giving support in the form of providing information on study tips that can help them to be successful in STEM majors. Previous research has also emphasized the significance of instructors of African American women who provide support as their students pursue STEM degrees in college (Brown, 2011; Fries-Britt & Holmes, 2012; Jordan, 2006; Perna et al., 2009; Taylor, 2015; Warren, 2000).

The final source of social capital reported by the African American women in STEM majors in this study is academic advisors. Academic advisors assisted these women with navigating through their coursework, for example, by providing them with information about course requirements (Ashford et al., 2017; Robinson & Franklin, 2011). Some African American women received positive and helpful advice from academic advisors regarding seeing instructors during office hours; registering for courses; filling out proper paperwork; and seeking recommendations for graduate school in math, biology, and engineering majors. For instance, Jennifer, a math major, explained, "my math advisors were really helpful. They said these are the different teachers teaching this class. These are teachers I think are really good. These are the teachers who I know have this history."

In a similar manner, when asked about how she learned about becoming successful in the biology major, Briana stated, "my academic advisors and then my one professor freshman year. He was really helpful. I would go to his office hours all of the time. He was by far the most helpful of all of my professors ever." These life stories point to the importance of academic advisors in promoting the success and advancement of African American women in STEM fields by providing them with information to advance in their respective majors. Earlier research has also demonstrated how academic advisors assist African American women in navigating through college (Perna et al., 2009; Robinson & Franklin, 2011; Young-Jones et al., 2013).

African American Women's Support Networks for Social Science and Health Science Majors in College

This study also found that despite adversity in biology and business majors, African American women who switched from these majors to sociology and health science majors also had parental, peer, church, and community support networks. Some African American women who left biology majors described support networks consisting of family members, peers, and church members. For instance, Carmen, a sociology major, spoke about parental support for her educational and career aspirations. She explained:

> My parents have always told me, 'You can do it. You are smart enough to be whatever you want to be. You just have to want it. You have to want to sacrifice for it.' That is a lesson that I am learning as I grow older.

Amber, a health science major, also had support from her parents and from her church family at home that helped her continue pursuing her undergraduate degree and career aspirations. When asked about support systems, she stated, "I would say my parents were my main people, and some of the pastors at church who I'm close with, mainly because they understand where I'm coming from."

Similarly, Crystal, a health science major, discussed support from her mother. "My mom has always been supportive, even when I told her I wasn't going to medical school anymore. She was like, 'I still support you.' So, my mom, I'd say, is my biggest supporter, always has been." Crystal also sought support from peers and church family members while she was in college. When Danielle experienced challenges in the health science and biology majors, she sought support from her mother as well. She vividly remembers her mother encouraging her to stick with biology. She recalled: "I'm always calling home. 'Mommy, I failed this test.' She was like, 'Change your study habits. Do the review.' She's my rock." Danielle had support from her older siblings as well. Consistent with scholarship on college persistence (Ashford et al., 2017; Committee on Underrepresented Groups and the Expansion of the Science and Engineering Workforce Pipeline et al., 2011; Tinto 2012), these stories provide evidence that support networks are important resources for the retention and graduation of African American women college students.

Conclusion

The purpose of this chapter was to better understand the high school and college support networks that help African American women prepare for STEM majors and aid in their college persistence in STEM, social science, and health science majors. For the African American women in this study, the use of collective resources provided them with social capital in the form of information and resources by showing them how to be successful in these majors at Town University. Successfully navigating these majors required that they find and access support networks consisting of peers, professors, family members, and academic advisors.

References

Anderson, D. L. (2019). *"Developing all these petals": A narrative study of the strategies and networks African American women at historically White institutions access, create, and employ to succeed* [Doctoral dissertation, University of Iowa]. Iowa Research Online.
Ashford, S. N., Wilson, J. A., King, N. S., & Nyachae, T. M. (2017). STEM SISTA spaces creating counterspaces for Black girls and women. In T. S. Ransaw and R. Majors (Eds.), *Emerging issues and trends in education*. Lansing: Michigan State University Press.
Banks, C. A. (2009). *Black women undergraduates, cultural capital, and college success*. New York: Peter Lang.
Borum, V. & Walker, E. (2011). Why Didn't I know? Black women mathematicians and their avenues of exposure to the doctorate. *Journal of Women and Minorities in Science and Engineering, 17*(4), 357-369. doi: 10.1615/JWomenMinorScienEng.2011003062
Britt, S. L., Younger, T. K., & Hall, W. D. (2010). Lessons from high-achieving students of color in physics. *New Directions for Institutional Research, 148*, 75-83. https://doi.org/10.1002/ir.363

Ceglie, R. (2013). Religion as a support factor for women of color pursuing science degrees: Implications for science teacher educators. *Journal of Science Teacher Education*, *24*, 37-65. https://doi.org/10.1007/s10972-012-9286-z

Chambers, C., & Wapole, M. (2017). Academic achievement among black sororities: Myth or reality? *College Student Affairs Journal*, *35* (2)131-139. doi:10.1353/csj.2017.0018

Coleman, J. S. (1990). *Foundations of social theory*. Cambridge: MA: The Belkmap Press.

Committee on Underrepresented Groups and the Expansion of the Science and Engineering Workforce Pipeline, Committee on Science, Engineering, and Public Policy, Policy and Global Affairs, National Academy of Sciences, National Academy of Engineering, & institute of Medicine. (2011). *Expanding underrepresented minority participation: America's science and technology talent at the crossroads*. D.C.: National Academies Press.

Conchas, G. Q. (2006). *The color of success: Race and high achieving urban youth*. New York: Teachers College Press.

Coneal, W. B. (2012). African American high-achieving girls: STEM careers as options. In C. R. Chambers & R. V. Sharpe (Eds.), *Black female undergraduates on campus: Successes and challenges (Diversity in Higher Education, Vol.12)* (pp. 161-183). Bingley, UK: Emerald Group Publishing Ltd.

Dixson, A. & Chambers. C. R. (2009). College predisposition and the dilemma of being black and female in high school. In V. B. Bush, C. R. Chambers, & M. Wapole (Eds.), *From diplomas to doctorates: The success of Black women in higher education and its implications for educational opportunities for all* (pp. 21-38). Sterling, VA: Stylus Publishing.

Falco, L. D. (2016). The school counselor and STEM career development. *Journal of Career Development*, 1-16. https://doi.org/10.1177/0894845316656445

Fries-Britt, S., & Holmes, K. (2012). Prepared and progressing: Black women in physics. In C. R. Chambers & R. V. Sharpe (Eds.), *Black female undergraduates on campus: Successes and challenges (Diversity in Higher Education, Vol. 12)* (pp. 199-218). Bingley, UK: Emerald Group Publishing Ltd.

Gold, S. P. (2011), Buried treasure: Community cultural wealth among Black American female students. In C. R. Chambers (Ed.), *Support systems and services for diverse populations: Considering the intersection of race, gender, and the needs of Black female undergraduates (Diversity in Higher Education, Volume 8)* (pp. 59-72). Bingley, UK: Emerald Group Publishing Limited.

Griffin, K. A., Pérez II., D., Holmes, A. P., & Mayo, C. (2010). Investing in the future: The importance of faculty mentoring in the development of students of color in STEM. *New Directions for Institutional Research*, *148*, 95-103. https://doi.org/10.1002/ir.365

Hanson, S. L. (2007). Success in science among young African American women: The role of minority families. *Journal of Family Issues*, *28*(1), 3-33. https://doi.org/10.1177/0192513X06292694

Hanson, S. L. (2009). *Swimming against the tide: African American girls and science education*. Philadelphia: Temple University Press.

Harding, S. (2006). *Science and social inequality: Feminist and poststructural issues*. Urbana: University of Illinois Press.

Harris, T. (2019). *Community cultural wealth brokers: A phenomenological study of the experiences of low-income, first-generation Black female undergraduates at a historically White institution* [Doctoral dissertation, Rutgers, The State University of New Jersey]. Rutgers University Community Repository.

Hurtado, S., Eagan, M. K., Tran, M. C., Newman, C. B., Chang, M. J., & Velasco, P. (2011). "We do science here": Underrepresented students' interactions with faculty in different college contexts. *Journal of Social Issues, 67*(3), 553-579. https://doi.org/10.1111/j.1540-4560.2011.01714.x

Ireland, D. T., Freeman, K.E., Winston-Proctor, C.E., DeLaine, K. D., Lowe, S. M., & Woodson, K. M. (2018). (Un)hidden figures: A synthesis of research examining the intersectional experiences of Black women and girls in stem education. *Review of Research in Education, 42*, 226–254. https://doi.org/10.3102/0091732X18759072

Jez, S. J. (2012). Analyzing the female advantage in college access among African Americans. In C. R. Chambers & R. V. Sharpe (Eds.), *Black female undergraduates on campus: Successes and challenges (Diversity in Higher Education, Vol.12)* (pp. 43- 57). Bingley, UK: Emerald Group Publishing Ltd.

Jordan, D. (2006). *Sisters in science: Conversations with Black women scientists on race, gender, and their passion for science*. West Lafayette, IN: Purdue University.

Koch, M., Lundh, P., & Harris, C. J. (2019). Investigating STEM support and persistence among urban teenage African American and Latina girls across settings. *Urban Education, 54*(2), 243-273. https://doi.org/10.1177/0042085915618708

Kuh, G. (2009a). What student affairs professionals need to know about student engagement? *Journal of College Student Development, 50*(6), 683-706. doi:10.1353/csd.0.0099.

Kuh, G. (2009b). *High-impact educational practices: What they are, who has access to them, and why they matter.* Washington D.C: Association of American Colleges & Universities.

Lee, O., & Luykx, A. (2006). *Science education and student diversity.* New York: Cambridge University Press.

Loury, G. (1989). Why should we care about group inequality? In S. Shulman & W. Darity II (Eds.), *The question of discrimination: Racial inequality in the U.S. labor market* (pp. 290). Middleton, CT: Wesleyan University Press.

McPherson, E. (2014). Informal learning in science, math, and engineering majors for African American undergraduates. *Global Education Review, 1*(4) 96-113.

McPherson, E. (2015). Having our say in higher education: African American women's stories of 'doing science' through using spiritual capital. In V. Evans-Winters and B. Love (Eds.), *Black feminism in education: Black women speak back, up, and out* (pp. 93-100). New York: Peter Lang.

McPherson, E. (2017). Oh you are smart: Young, gifted, African American women in STEM majors. *Journal of Women and Minorities in Science and Engineering, 23*(1), 1-14. doi: 10.1615/JWOMENMINORSCIENENG.2016013400

Moore, J. S. (2016). *Do I belong? Narratives of sense of belonging and fit from underrepresented African American and Latina women in science undergraduate majors* [Doctoral dissertation, Rowan University]. Rowan Digital Works.

Morton, T., & Parsons, E. (2018). #BlackGirlMagic: The identity conceptualization of Black women in undergraduate STEM education. *Science Education, 102*(6),1363-1393. https://doi.org/10.1002/sce.21477

Parker, A. D. (2013). *Family matters familial support and science identity formation for African American female scholars* [Doctoral dissertation, University of North Carolina- Charlotte]. ProQuest Dissertations and Theses Global.

Perna, L., Lundy-Wagner, V., Drezner, N.D., Gasman, M., Yoon, S., Bose, E., & Gary, S. (2009). The contribution of HBCUs to the preparation of African American women for STEM careers: A case study. *Research in Higher Education, 50,* 1-23. doi: 10.1007/s11162-008-9110-y

Robinson, S., & Franklin, V. (2011). Working against the odds: The undergraduate support needs of African American women. In C. R. Chambers (Ed.), *Support systems and services for diverse populations: Considering the intersection of race, gender, and the needs of Black female undergraduates* (Diversity in Higher Education, Volume 8) (pp. 21-41). Bingley, UK: Emerald Group Publishing Limited.

Simmons II, R. W. (2013). It can be done and it must be done: Creating educational excellence for African American girls in urban science classrooms. In E. Zamani-Gallaher & V. C. Polite (Eds.), *African American females: Addressing challenges and nurturing the future* (29-44). East Lansing: Michigan State University Press.

Taylor, W. A. (2015). *Counterstories: Educational resilience of adult African American women attending an urban predominantly White university.* [Doctoral dissertation, University of Louisville]. The University of Louisville's Institutional Repository.

Tinto, V. (2012). *Completing college: Rethinking institutional action*. Chicago: University of Chicago Press.

Wapole, M. (2009). African American women at highly selective colleges: How African American campus communities shape experiences. In V. B. Bush, C. R. Chambers, & M. Wapole (Eds.), *From diplomas to doctorates: The success of Black women in higher education and its implications for educational opportunities for all* (pp. 85-107). Sterling, VA: Stylus Publishing.

Warren, W. (2000). *Black women scientists in the United States*. Bloomington: Indiana University Press.

Williams, P. C. (2011). Beating the odds: How five non-traditional Black female students succeeded at an Ivy league institution. In C. R. Chambers (Ed.), *Support systems and services for diverse populations: Considering the intersection of race, gender, and the needs of Black female undergraduates (Diversity in Higher Education, Volume 8)* (pp. 89-106). Bingley, UK: Emerald Group Publishing Limited.

Winkle-Wagner, R. (2009). An asset or an obstacle: The power of peers in African American women's college transition. In V. B. Bush, C. R. Chambers, & M. Wapole (Eds.), *From diplomas to doctorates: The success of Black women in higher education and its implications for educational opportunities for all* (pp. 55-71). Sterling, VA: Stylus Publishing.

Yosso, T. J. (2005). Whose culture has capital? A critical race theory discussion of community cultural wealth. *Race Ethnicity and Education*, *8*(1), 69-91. https://doi.org/10.1080/1361332052000341006

Young-Jones, A. D., Burt, T. D., Dixon, S., & Hawthorne, M. J. (2013). Academic advising: Does it really impact student success? *Quality Assurance in Education*, *22*(1), 7-19. https://doi.org/10.1108/09684881311293034

Young, J. L., Feille, K. K., & Young, J. R. (2017). Black girls as learners and doers of science: A single-group summary of elementary science achievement. *Journal of Science Education, 21*(2) 1-20.

Young, J. L., Young, J.R., Ford, D. Y. (2017). Standing in the gaps: Examining the effects of early gifted education on Black girl achievement in STEM. *Journal of Advanced Academics, 28*(4), 290-312. https://doi.org/10.1177/1932202X17730549

Chapter 6

Lost Talent in the Leaky Pipeline: African American Women Cooled-Out From STEM Majors

Chapter 6 asks the question "How does the cooling-out process within the science, technology, engineering, math (STEM) culture function to push African American women out of college STEM majors at a predominantly White institution (PWI)?" African American women in college frequently leave STEM majors because they are pushed out during the cooling-out process by the departmental climate (e.g., curriculum, peers, professors, academic advisors) in higher education settings. This chapter begins by discussing the sociological concept of cooling-out, which was first applied to higher education by Burton Clark in 1960. It then applies the cooling-out concept to understanding the process involved in the departure of African American women from STEM majors.

Cooling-Out

Cooling-out is a process in higher education in which community college students are discouraged from pursuing their educational aspirations (Clark, 1960, 1980), causing them to seek out alternatives. The cooling-out process consists of five phases. Phase 1 comprises of alternative achievement, in which

> substitute avenues may be made to appear not too different from what is given up, particularly as to status… One does not fail but rectifies a mistake. The substitute

status reflects less unfavorably on personal capacity than does being dismissed and forced to leave the scene. (Clark, 1960, pp. 574-575)

The second phase, gradual disengagement involves:

> a gradual series of steps, [in which] movement to a goal may be stalled, self-assessment encouraged, and evidence produced of performance. This leads toward the available alternatives at little cost...Compared with the original hopes, however, it is a deteriorating situation. If the individual does not give up peacefully, he [or she] will be in trouble. (Clark, 1960, p. 575)

The third phase, objective denial, focuses on reorientation based on facts. This might include "a record of poor performance [that] helps to detach the organization and its agents from the emotional aspects of the cooling-out work" (Clark, 1960, p. 575). The fourth phase, consolation, centers on the use of counseling. As Clark (1960) states, "counselors are available who are patient with the overambitious and who work to change their intentions. They believe in the value of the alternative careers, though of lower social status, and are practiced in consoling" (p. 575). In the final phase, avoidance of standards,

> the cooling-out process avoids appealing to standards that are ambiguous to begin with. While a "hard" attitude toward failure generally allows a single set of criteria, a "soft" treatment [as provided in the consolation phase] assumes that many kinds of ability are valuable, each in its place, [with students achieving] proper classification and placement [in their alternative majors]. (Clark, 1960, p. 575)

Typically, the cooling-out framework has been applied to community college settings to account for why some students change majors to earn terminal degrees rather than obtaining a degree in their initial college aspiration (Alexander et al., 2008). Few scholars have studied the cooling-out framework in four-year college settings (Baird, 2014). This chapter applies the cooling-out framework to understand how African American women are pushed out of their STEM majors at Town University, a public four-year institution.

Alternative Achievement

The African American women in this study encountered alternative achievement in their college classrooms through the academic rigor of gatekeeper courses in the sciences. Crystal, a health science major, stated, "I dropped Molecules and Cells both times that I took it. The first time, I may have not worked as hard as I thought. It was freshman year. I was used to not getting a challenge in science classes." Danielle, another health science major, further explained the teaching practices within biology gatekeeper classes at Town University. She pointed out:

> I feel like some classes are definitely weeding-out classes. So many people have changed their majors. They say, 'I can't do this.' And then they'll go into health sciences or something completely easy, and I'm just like, if you just push through, this was not that hard.

Similarly, health science major Raven acknowledged the weed-out process in science classes. She recalled:

> In General Chemistry I, they started weeding people out, and people started saying 'Well, maybe this just isn't for me.' So as soon as you get up, you realize how many people have dropped off from when you were

in General Chemistry I and Evolution and Organisms, Molecules and Cells. For General Chemistry I, there were probably 400 of us. Maybe about 40 or 50 were Black. Now, in my Biochemistry class, it's about the same amount. It's probably 10 Black students.

These narratives indicate that the gatekeeper courses in the sciences at Town University made African American women call into question their STEM major status. Thus, gatekeeper courses in the sciences and the academic rigor of science courses accounted for the departure of some African American women from STEM majors. This finding supports prior research (Espinosa, 2011; Malcom & Malcom, 2011; McCoy et al., 2017; Ong et al., 2011) that there is an underrepresentation of women of color in STEM majors as a result of gatekeeper courses.

Within the cooling-out framework, in terms of alternative achievement, the African American women in this study made mistakes in terms of failing to build relationships with their White peers. In fact, they felt disconnected from their chilly White peers in biology majors and chemistry majors at Town University. For instance, a chemistry major and former health science major, Kayla described difficulties in working with peers in study groups for physics, biology, and chemistry classes. She pointed out:

> Usually, there's not a lot of minorities in those classes, and you have to invite yourself to the groups or ask questions. They won't come to you. Even in group work, I can tell there's a kind of dynamic to it. You still have to prove your presence. At first, it was kind of annoying, but I just got used to it. If you make your presence known in the beginning, I feel like it's easier as the semester progresses.

Similarly, when asked about experiences working with science peers, Raven, a health science major, explained,

> My lab partner was White. He was pretty much to himself. He only talked if he needed a measurement from me. He was really standoffish. I don't know what that was about, but it did make a difference, because that was the only lab that I got a C in, so I don't know if it was the class or how we interacted or what affected my grade, because the only grades that we really had were the test and the lab reports.

These narratives suggest that a failure to build relationships with White peers contributed to some African American women being cooled-out from STEM majors at Town University. As a result, these women were more likely to lose their STEM major status. Such experiences are consistent with the research on cooling-out, which provides evidence that students are subject to losing their STEM major status during the cooling-out process in higher education (Clark, 1960, 1980; Lanell, 2017; Moore, 1975).

Gradual Disengagement

After experiencing course failure and an inability to work with peers, some African American women began to slowly disengage from their STEM majors. For instance, Crystal, a former biology major who switched to health science, described her disengagement in a biology class at Town University as follows:

> I honestly just probably gave up. That was probably when I realized that biology was not going to work. The first time I probably just gave up because I was not about to get an F on my transcript and my mom is going to have a fit. I am probably going to have to come home.

Carmen, who switched her major to sociology, was very emotional when describing her painful experiences deciding whether to remain in or leave her original major, biology. She eloquently stated:

> When I first got to Town University, I thought I am going to take responsibility for my actions. However, I was not as focused as I needed to be. I failed drastically. I failed myself. I failed my family. I should have been better and I wasn't better. I was getting the worst scores that I have ever gotten in my life. Of course, you know the first semester I got put on probation and I know why. I was crying every day. I had never failed anything in my entire life, so I was shell-shocked. What I used to feel for science, the excitement that I used to feel for science or playing doctor in the field waned. Or the interest and intrigue in the field that I had just sitting and looking into my retina, looking at the different levels of the eyeball, seeing the grass; where did these berries come from? Now that I reflect on my experiences, it is not that I don't enjoy it, but I feel more so withdrawn from it. I did not feel this way until I got here. It is something about Town University that just kills you, your soul, your being—at least for me, because I went through so much here with my science experiences.

These narratives point to the cooling-out function of higher education within the culture of science, which prompted several of the African American women in this study to disengage from their STEM courses.

In addition to course disengagement, African American women experienced the coldness of professors in terms of their teaching styles, approachability, and availability. Consider the case of Carmen, a sociology major and former biology major, who struggled in science classes taught by White male professors at Town University. She recalled: "I tried to build a bond with my instructors. I really wish that I had a relationship with my instructors and my peers at Town University, because it seems like that was the thing I was missing." Additionally, a health science major and former biology major, Crystal described her frustration with her lack of connection with her science professors. She stated:

> I really think that my professors did not fully comprehend that I did not fully understand the class materials. It was times when I went to office hours, and I didn't know where to begin to ask them a question. I don't know what I do not understand at the beginning and it was like professors thought I did not put time in. After those experiences, I thought, 'I cannot communicate with you what I do not understand, because I don't know what I am supposed to understand.' I just don't think that the professors quite understand that in science classes.

These accounts make clear that some of the African American women in this study encountered a chilly environment during their instructors' office hours that hindered their achieving a solid grasp of the course materials. As a result, they disengaged from their science courses. They also had to adjust to the teaching styles and curriculum of the dominant White male classroom context. Consistent with scholarship on the chilly climate for women in STEM majors (Espinosa, 2011; Fries-Britt & Holmes, 2012; Hughes, 2014; McCoy et al., 2017; McPherson, 2017; Ong et al., 2011; Sosnowski, 2002), the unwelcoming, chilly environment of their White male professors might have contributed to African American women's disengagement and later departure from biology majors at Town University as well.

Moreover, peers created a chilly environment for some of the African American women in the study, which played a role in their disengagement from STEM majors. For example, Crystal, a health science major and former biology major, said:

> I feel like some people did not care. I also feel like some people thought, 'Oh, it is an African American woman in the science class doing pre-med.' They said, 'You are not going to go all of the way.' That was one reason that I was so scared to leave the pre-med track. I felt that, in science classes, nobody was really anxious to help you. I don't know if it was that they did not understand it or they did not want to help. I would not ask others for help in science here. It is just like they are not going to want to help. I just sat there continuing to be confused in class.

Likewise, Raven, also a health science major and former biology major, reflected on a negative peer experience that she had in a biology class. She pointed out:

> In Genetics, we had a project that we had to do in sequencing a certain insect. We were supposed to be working on the project. I had an Asian partner. I had been trying to contact her all week, and she kept making excuses, and then she showed up the night before the assignment was due with barely anything done, and my stuff was done already. So, then, I had to basically explain the entire project to her, because she hadn't done anything. I sent it back to her, and she was like, 'Okay, that's fine.' I got there to present the next day, and she didn't show up. So, I had to do the project by myself. And granted, most of it was mine that I had already done, so it wasn't a problem with

me doing it, but that was a stressful period of trying to communicate and half the reason why I don't like group projects now.

These narratives suggest that the culture of science promotes isolation for African American women by their peers, as has been shown in prior studies (Brown, 2011; Bryant, 2019; Dortch & Patel, 2017; Fries-Britt & Holmes, 2012; Ireland et al., 2018; Jordan, 2006), and thus cools out African American women from STEM majors through disengagement. Within this culture at Town University, as a result of their interactions with their peers, several African American women felt like misfits who did not belong in chemistry or biology majors. The culture of science also pushed them to gradually disengage from their departments because of the chilly climate of professors and peers, which is consistent with the cooling-out function of higher education (Clark, 1960, 1980; Carroll et al., 2016; Larnell, 2017; Moore, 1975).

Objective Denial

The African American women in this study also exhibited denial about leaving their STEM majors, even though they performed poorly in their science classes. A former biology major and current sociology major, Carmen vividly remembered the academic rigor of some of the science classes she took at Town University. She recalled flunking two science classes at Town University. However, the academic rigor was different at the local community college. She stated, "last summer, I took the classes over at Town Community College. I got A's in them. I am just glad that is over…I just feel like they got me in to fill a quota and then push me out."

When asked about the time when she wanted to depart her health science major, former biology major Amber stated:

> I probably feel like I kind of want to leave the biology major. That's only because I keep failing the exams, and I'm just thinking, "Is this really for me? Is this really what God wants me to do with my life?" At the same time, I want to be a doctor, so I have to know this information, so I'm going to stick it out.

These descriptions illustrate that some African American women were in denial about their performance in their STEM majors and thus wanted to remain despite the adversity that they faced. As reported in previous research (Espinosa, 2011; Ong et al., 2011; McCoy et al., 2017; Stewart et al., 2008; Weston et al., 2019), the academic rigor and course failure encountered in weed-out courses contribute to some African American women students thinking about leaving undergraduate STEM majors. These findings are also consistent with the objective denial phase within the cooling-out function in higher education (Clark, 1960, 1980; Moore, 1975).

Agents of Consolation

Several African American women in STEM majors at Town University encountered the chilly culture of science through their White male academic advisors, who counseled them to leave their STEM majors. Consider for example, the experience of Kayla who switched her major from health science to chemistry. She explained:

> I went to my White male chemistry advisor's office hours once, and he actually told me not to do the major. He asked me if I was a fan of math, because there's a

lot of math that goes into it. He wasn't really good at math, so that kind of changed his perspective. So, I'm just like, "What are you trying to say?" He was just trying to tell me not to do it. So, I'm like, "Whether I like math or not, I'm going to major in this."

Kayla's story suggests that a White male chemistry academic advisor tried to dissuade her from pursuing a chemistry major. When asked about her feelings during this advising meeting, Kayla stated "I had this look on my face. I know I could have been disrespectful right then, so I tried to stay silent. So, I didn't really react to it, I just walked out. I'm not going back to him."

Carmen's biology academic advisor, also a White man, literally pushed her out of the biology major. When asked about the types of referrals given by biology advisors, Carmen, a former biology major, stated, 'drop out of school, take some time off, stop-out, change your major.' When further probed about her feelings after the conversation with her advisor, she recalled, "I hated them. I hated everybody that I talked to at one point. That is why I don't talk to them anymore."

These stories indicate that White male academic advisors at Town University discouraged some of the African American women who participated in this study from pursuing science majors, because of their perceived poor fit in STEM majors. This experience is consistent with the consolation phase within the cooling-out process of higher education (Clark, 1960, 1980; Moore, 1975).

The chilly academic advising environment within the culture of science at Town University might explain why some African American women were cooled-out of science majors and chose majors with warmer advising climates. After leaving the biology major, several African American women described their academic experiences in social science or health science majors in highly positive terms. For example, Carmen who switched to sociology, found the new major to be more suited to her abilities. She stated:

> Changing majors helped because when I changed my major to sociology, I can do in my sleep. I am not saying this to be cocky, but to be low-key. As many points as my GPA [grade point average] dropped from failing in the biology major, I have made up plus more just by taking sociology courses in two semesters, because I have gotten straight A's. Just because the level of rigor is not the same.

Raven, who changed her major to health science, reported:

> I just transferred in my junior year, and I've learned a great deal that I probably wouldn't have learned by just staying in biology. My new department has given me more opportunities that are not necessarily about the sciences, but that could direct me into the extras that you need to be well-rounded. Everyone is really open to helping each other and helping you and pointing you in the direction to know things. My advisor is great. She's always sending me things; even if I'm not in the office, she'll send me things over email: 'Hey, I found this. Check this out.' So as far as welcoming and helping and learning about different opportunities, I love health sciences. The department is really great.

Crystal, who also switched to health sciences, agreed that the new department was welcoming. She stated:

> We have an office in the Department of Health Sciences where you go and you schedule an appointment. They greet you by saying, 'Hey, how are you doing?' You see a TA [teaching assistant] in the hallway or

professors, and it is like, 'Hey!' It might be a smile. It may not be a hello. It is just a friendly environment.

In addition, another new health science major, Danielle, described feeling welcomed in her new department. She noted:

> They tell me what classes to take. I know my advisor for the Health Sciences program, and I love her. She just likes to know about your life, and she really cares to know. My mom had surgery last Thursday, and my advisor said she'd be praying for her. Well, you know, Thursday came up, and my advisor asked, 'How's your mom doing?' It was really nice to have someone who's trying to be involved in your life.

These statements show that switching out of the biology major promoted academic confidence and redirected these African American women to the end goal of college graduation. Their stories also demonstrate that, although they were cooled-out of STEM majors, they were able to find warmer climates in health science and sociology majors. This type of experience is similar to findings from community college research that revealed a reduction in educational degree aspirations over a period of time as a result of the cooling-out process (Broton, 2019, Clark, 1960, 1980; Moore, 1975).

Avoidance of Standards

African American women observed that Town University avoided standards by failing to provide equal opportunities for African American women to study STEM majors. Consider the lived experiences of Regina, a sociology major and pre-business major. She noted:

> The opportunities and the competition are so stiff. It is almost discouraging. That is why I don't really

like to talk about it, I know that I have to be realistic, that is me. Yeah, I want to be optimistic and I want to be encouraging, but it is really real, I cannot ignore that.

Similarly, Carmen, a sociology major and former biology major explained:

> I think that everybody experiences their challenges at one point in time or another. I don't think that is distinctive to one group. Women struggle, because there are not a lot of us. Black women struggle, because there are not a lot of us. Black men struggle, because there are even less of them. Everybody has a problem here or there because we have institutionalized racism and segregation still. Even though I experienced racism, it doesn't hold me captive. I don't participate in it. I don't promote it. I try to live beyond it. Become familiar with your situation and work within it.

These responses indicate that not all African American women have an opportunity to study STEM majors at Town University. To avoid the standards of engagement in STEM majors, some African American women were cooled-out by being redirected to health science or social science majors. In accordance with the scholarship on cooling-out in community colleges (Clark, 1960, 1980), Town University was never held accountable for failing to provide equal opportunities for African American women who were pursuing STEM majors. Instead, Town University hid these cooling-out practices from the African American women so that they would be able to graduate from college in other fields rather than encouraging the women to stick with the STEM majors to pursue their career aspirations.

Conclusion

The experiences reported in this section provide evidence that the culture of science can produce feelings of self-doubt, defeat, and a loss of academic confidence among African American women. The culture of science also promotes academic failure and the retaking of gatekeeper courses. The gatekeeper courses, isolation, and a chilly culture can help explain why some African American women were cooled-out from STEM majors and transferred from biology majors to social science or health science majors at Town University. This chapter thus supports the conclusion that there is a cooling-out process in higher education that results in fewer African American women pursuing STEM majors. As a result, fewer African American women are in a position to go into careers in the STEM fields or become medical doctors.

References

Alexander, K., Bozick, R. & Entwisle, D. (2008). Warming up, cooling out, or holding steady? Persistence and change in educational expectations after high school. *Sociology of Education*, *81*, 371-396. https://www.jstor.org/stable/20452746

Baird, A. (2014). *The social function of for-profit higher education in the United States* [Doctoral dissertation, University of Central Florida]. UCF Electronic Theses and Dissertations.

Broton, K. M. (2019). Rethinking the cooling out hypothesis for the 21st century: The impact of financial aid on students' educational goals. *Community College Review*, *47*(1), 79- 104. https://doi.org/10.1177/0091552118820449

Brown, J. (2011). *African American women chemists*. New York: Oxford University Press.

Bryant, T. (2019). *Unhidden and unrelenting figures: The persistence of Black women in STEM disciplines* [Doctoral dissertation, California State University]. ProQuest Dissertation and Theses Global.

Carroll, J.M., Mueller, C. & Pattison, E. (2016). Cooling out undergraduates with health impairments: The freshman experience. *The Journal of Higher Education*, *87*(6), 771- 800. doi: 10.1353/jhe.2016.0029

Clark, B. R. (1960). The cooling-out function in higher education. *The American Journal of Sociology*, *65*(6), 569-576.

Clark, B. (1980). The "cooling-out" function revisited. In G. Vaughn (Ed.), Questioning the community college role. *New Directions for Community College*, *32*, 15-31.

Dortch, D. & Patel. C. (2017). Black undergraduate women and their sense of belonging in stem at predominantly white institutions. *NASPA Journal About Women in Higher Education*, *10*(2), 202-215. https://doi.org/10.1080/19407882.2017.1331854

Espinosa, L.L. (2011). Pipelines and pathways: Women of color in undergraduate STEM majors and the college experiences that contribute to persistence. *Harvard Education Review*, *81*(2), 209-240. https://doi.org/10.17763/haer.81.2.92315ww157656k3u

Fries-Britt, S., & Holmes, K. (2012). Prepared and progressing: Black women in physics. In C. R. Chambers & R. V. Sharpe (Eds.), *Black female undergraduates on campus: Successes and challenges (Diversity in Higher Education, Vol. 12)* (pp. 199-218). Bingley, UK: Emerald Group Publishing Ltd.

Hughes, R. (2014). The evolution of the chilly climate for women in science. In J. Koch, B. Polnick, & B. Irby

(Eds.), *Girls and women in STEM fields: A never-ending story* (pp 71- 92). Charlotte, NC: Information Age Publishing.

Ireland, D. T., Freeman, K.E., Winston-Proctor, C.E., DeLaine, K. D., Lowe, S. M., & Woodson, K. M. (2018). (Un)hidden figures: A synthesis of research examining the intersectional experiences of Black women and girls in stem education. *Review of Research in Education*, *42*, 226–254. https://doi.org/10.3102/0091732X18759072

Jordan, D. (2006). *Sisters in science: Conversations with Black women scientists on race, gender, and their passion for science*. West Lafayette, IN: Purdue University.

Larnell, G. V. (2017). On the entanglement of mathematics remediation, gatekeeping, and the cooling-out phenomenon in education [Paper presentation]. International Mathematics Education and Society 9th Annual Conference, Volos Greece.

Malcom, L., & Malcom, S. (2011). The double bind: The next generation. *Harvard Educational Review*, *81*(2), 162-172. https://doi.org/10.17763/haer.81.2.a84201x508406327

McCoy, D. L., Luedke, C. L., & Winkle-Wagner, R. (2017). Encouraged or weeded out: Perspectives of students of color in the STEM disciplines on faculty interactions. *Journal of College Student Development*, *58*(5), 657-673. doi:10.1353/csd.2017.0052.

McPherson, E. (2017). Oh you are smart: Young, gifted, African American women in STEM majors. *Journal of Women and Minorities in Science and Engineering*, *23*(1), 1-14. doi: 10.1615/JWomenMinorScienEng.2016013400

Moore, K. M. (1975). The cooling out of two-year college Women. *Personnel and Guidance Journal*, *53* (8), 578-583.

Ong, M., Wright, C., Espinosa, L. L., & Orfield, G. (2011). Inside the double bind: A synthesis of empirical research on undergraduate and graduate women of color in science, technology, engineering, and mathematics. *Harvard Educational Review, 81*(2), 172-208. https://doi.org/10.17763/haer.81.2.t022245n7x4752v2

Sosnowski, N. (2002). *Women of color staking a claim for cyber domain: Unpacking the racial/gender gap in science, mathematics, engineering and technology* [Doctoral dissertation, University of Massachusetts]. ProQuest Dissertations and Theses Global.

Stewart, G., Wright, D., Perry, T., & Rankin, C. (2008). Historically Black Colleges and Universities: Caretakers and precious treasures. *Journal of College Admission, 201*: 24- 29.

Weston, T. J. Seymour, E. Koch, A. K. & Drake, B.M. (2019). Weed-out classes and their consequences. In E. Seymour & A. Hunter (Eds.), *Talking about leaving revisited: Persistence, relocation, and loss in undergraduate education* (pp. 197-243). Switzerland: Springer.

Chapter 7

Dream On!: African American Women's Thoughts on College Completion

This chapter explores African American women's college completion dreams in science, technology, engineering, and math (STEM), social science, and health science majors. It shows that African American women who remained in or departed from STEM majors had dreams of completing college for personal, educational, or career goals.

College Completion Dreams of STEM Majors

In STEM majors, African American women in this study had dreams of completing college for both personal reasons and career preparation. For instance, a biology major, Briana discussed her reasons for wanting to graduate from college as follows:

> For personal reasons number one. I know that my parents would be proud of me. Two, just to have bragging rights; just to say that I'm an educated Black woman and I have a degree. Three, to get a good job, a job that I enjoy.

Ashley, another biology major, said,

> Probably one of the reasons I want to complete my degree is that I find it interesting, not just biology but all sciences. Also, there are definitely not a lot of Black people who graduate with degrees in science and then decide to pursue that in graduate school or do research,

so I don't feel obligated to do it, but it is definitely something that I look forward to doing.

Personal reasons also motivated Rachelle, who explained that she wanted to complete her biology major "because, going into college, a lot of my peers told me that I would not stay in it." Rachelle went on to say that:

> "A lot of Town University graduates would say 'I was in biology and then I quit.' Or 'No one stays in biology.' I just wanted to prove that I could do it. A lot of my friends dropped out of it as well."

Lara, a math major added:

> I just want to graduate with a degree. That is my main thing. I just entered with math because, at the time, I was ahead. I have always liked math as well. I feel like, when I came to college, if I was not going to major in math then I was going to major in chemistry.

Engineering major Simone mentioned a combination of financial, personal satisfaction, and career goals. When asked why she wanted to complete college, she said, "One, to support myself. Two, to feel like I completed it and made it. Three, I kind of want to work in the career and make money."

Other African American women desired to complete their STEM majors primarily to prepare for their future careers. For example, biology major Shannon explained, "pretty much the main reason that I majored in biology is because that is the major to prepare me most for medical school. That is my ultimate goal. That was why I chose this major in particular." Business major Patricia also focused on career goals. She explained:

> One, I know there are a lot of opportunities out there for me in business. Two, coming from Town University

with a business degree is just so much better in terms of prestige. Three, I just want to set an example for my family just by graduating period, but also graduating, having a job, or attending to all of these different opportunities that I've had thus far and what I can potentially have in the future.

Chemistry major Kayla identified a career goal as driving her selection of a STEM major but was also motivated by personal reasons. She responded:

One reason for majoring in chemistry is career wise, because I was thinking about pharmacy. Pharmacy is more related to chemistry than anything else. Second, I would be just graduating in general as an example for my siblings, being a first-generation college student and taking on a hard major like the hard sciences and knowing that there are no limits. Three, I just like chemistry a lot. I would rather do chemistry than anything else.

These personal reflections suggest that it is important for African American women to continue to aspire to complete college for personal and career reasons. By dreaming, they are in fact setting goals to commit to graduating from college.

College Completion Dreams of Social Science and Health Science Majors

Many of the African American women in social science and health science majors at Town University also aspired to complete college primarily to further their educational and career goals. For example, Regina, a sociology major who had started in pre-business, listed a variety of reasons for desiring to earn her degree. She said: "One, I guess because it goes nicely with public health. Two, it offers a lot of insight into group interactions that I am finding to be true. Three, it is flexible in terms of career possibilities"

Carmen, who switched her major from biology to sociology, explained her motivation for both majors in terms of her desire to become a doctor. She explained:

> With biology, I think the one reason I wanted to complete my degree would be because they said that I couldn't. That is really the only one. Honestly, I can be a doctor with any degree. I know people who have business degrees, English degrees, political science degrees, even degrees in sociology who are doctors now. My reason for majoring in biology used to be because I had to have this type of curriculum to get into medical school. As I grow older and the more curriculum that I take, I realize that is not true. So, now it is just because they said that I couldn't. For sociology, I want to complete my major because I started it. I need to finish it. You know honestly, I am starting to embrace it. It sets me apart from other medical school applicants.

Similarly, Amber, who switched from biology to health science, remained focused on completing her major so that she could attend medical school. She elaborated:

> A health science major will give me more options to do what I want to do if I go to medical school right away or not. Health science is broader, and I can do more and I can add more to it than with biology. To add more to biology, I have to actually do something totally different in order to add to it. Health science is broad so I can start out in it and then narrow down, whereas biology is narrow, so it's going to be harder to broaden out.

Raven, who also switched from biology to health science, offered very similar motivation for completing her new major. She stated:

> I started off in biology and I switched, because I felt that biology only gave me the technical part of medicine like the biology of knowing the different mechanisms of the body. Health science gives you the whole scope of the health world and health care. Health science also requires you to have an internship in order to graduate, which is something that biology doesn't offer. I realized that I got the experience that I needed outside of the classroom. Health science allows me to explore the health care world a little bit more. It has kind of confirmed what I want to do. Whereas biology, I knew what they taught me. With health science you get all types of health care careers, and I got to experience all of them. I knew that was not what I wanted to do. I knew that medicine is the route that I want to take.

Other African American women who switched from STEM majors at Town University wanted to graduate for more personal reasons. For example, Danielle provided a highly personal reason for completing her health science major, after switching from biology. She said, "I feel like this is definitely the path God has chosen for me, even being at this school, a school that I really didn't want to go to." Another former biology major who switched to health science, Crystal, also reported having personal motivations for completing her degree. She reported:

> One, because health has always been something I've always been interested in, whether it be a doctor or a nurse, just healthcare is a big issue to me. I personally

know people who have had issues getting healthcare, and that's my way of giving back. I find that I am passionate about it. Two, I want to get a job that I'm happy going to. This may sound cliché; I don't want to wake up and dread going to work every day. I want to wake up and love going to work, love what I do, love the people I work with. Three, I want to make my family proud and myself too. In my family, I have a lot of cousins. I think maybe three or four of us have gone to college. One before me has graduated. So, I'll be the second to graduate, and then there are some cousins behind me. Also, I know it's a big deal to my mom, and it's a huge deal to my grandmother.

Celeste, who changed her major from pre-business to sociology, was also motivated by personal and altruistic reasons to complete her new major. She stated:

Towards the end of my studies, I found a lot of problems with society. I guess my biggest reason to complete my major would be to have some type of social justice perspective one day in the future. Another reason would be just because I need to graduate college because I'm the first person in my family to go to college, so that would make me a family historian and trailblazer.

These stories illustrate the personal and career reasons the study participants had to finish college. By pursuing their dreams, they put themselves into a position to graduate from college.

Conclusion

All of the African American women in this study had dreams and aspirations for graduating from college. These desires were based on personal, educational, and career aspirations. To retain African American women in STEM, social science, and health science majors, it will be important for administrators, faculty members, and mentors to facilitate the success of these students through goal-setting and encouragement so that they can pursue their personal, educational, and career goals and dreams.

Chapter 8

Recommendations for STEM Student Success for African American Girls and Women

This chapter provides recommendations for instructors and other educational practitioners who work with African American girls and women in science, technology, engineering, and math (STEM). The chapter begins with a discussion of how to cultivate the science and math identities of African American girls in K–12 settings, followed by tips on how to engage African American girls in STEM in elementary and secondary schools. Next, recommendations are provided for how to more effectively teach African American girls in science and math classes by understanding their learning styles. The importance of exposing African American women to the hidden curriculum in postsecondary settings and establishing student support networks for African American girls and women in STEM is also discussed. Finally, recommendations are made to provide warmer departmental climates to facilitate the college graduation of African American women in STEM majors.

Cultivate Science and Math Identities in K-12 Settings

To cultivate African American girls' science and math identities, teachers should consider employing science and math projects as early as elementary school. Science and math identity formation will then continue through engagement in science and math projects during the adolescent years in middle and high school.

Engage African American Girls in STEM in K-16 Settings

To engage African American girls in science and math, K-12 teachers should use caring instruction (Siddle Walker, 2001) and culturally responsive teaching (Gay, 2018) to cultivate their learning. Using a hands-on approach and asking African American girls to serve as group leaders are very important for engaging these students in science and math in elementary, middle, and high school settings. In fact, in this study, I provided evidence that African American girls and women have been found to learn best in interactive environments that use culturally responsive teaching approaches for science and math courses.

From the lived experiences of the participants in this study, African American girls are engaged in math and science in elementary, middle, and high schools. College is when African American women are not actively engaged in the classroom in math and science classes. Consider the comments of Briana, a biology major. She stated: "I am usually not at all engaged in what my professors are talking about. The only reason why I do well in the math classes is because it is intuitive. It is logical. It makes sense." For the specific example of Introduction to Statistics, she added:

> I don't even go to class. I am not even going to lie. My professor posts the notes from that entire week, because we only have lectures twice a week. She posts the notes from the whole week every weekend, so I just copy the notes at my leisure. The concepts are so basic, because they are math based. You do all of the calculations. I am good with math, so I am acing all of the homework. I am getting B+'s and A's on my exams. I am like, "Hey, this is good!" It is not good

that I don't go to class, but it is good that I am still doing well in the class.

Carmen, a former biology major who switched to sociology, agreed by stating: "Honestly, for statistics, I never went to class. I only went to the first two classes." She also described her disengagement from a math class during her time attending a Historically Black College and University (HBCU). She explained:

> In my Calculus III course at Brownstone University [an HBCU], I died slowly every day. It was like Swahili. I did not know what the professor was talking about at all. Brownstone University has smaller courses, because it is a smaller university. It is like a lecture. You just go to lecture. It is two hours. The first half is lecture; the second half, like 30 minutes, is asking questions, and you got the rest of the time to do your homework.

Shannon, a biology major, was sometimes engaged in her math classes, depending on the situation. She explained:

> In math, sometimes if I have a question to ask right then, like when a professor is in that example or explaining a certain step, I will speak up. So pretty much for clarification or if the professor needed participation and I knew the answer, I would participate. In precalculus at Town University, no I did not participate. In calculus, yes, I participated because it was a smaller classroom at the junior college. The calculus professor did have a couple of times when he would have students come up to the front of the class and write their solutions to certain problems.

These reflections provide evidence that African American women in this study were withdrawn from math classrooms at the college level.

Similarly, African American women were disengaged in their college science classrooms. They rarely asked questions in class because of feelings of intimidation, as reported by biology major Briana:

> To raise my hand and ask a question is very intimidating, because it is a big room full of people. I don't like speaking in front of people, big groups of people. I will just wait to go to office hours. It is not that I am too intimidated in terms of asking the professor. It is just that, around large groups of people, it is very intimidating. When I do ask a question during class, it is usually because I am engaged in what the professor is talking about, or because I have a question and I really need him to explain it before he moves on. I have gotten more comfortable asking questions in lecture over the years, only because I feel like I have gotten more open about that. Freshman year, I did not do it at all. I didn't feel comfortable with that. I felt kind of lost. This professor isn't gonna remember me. He doesn't care.

Shannon, another biology major, also expressed her reluctance to actively engage in science classes. She remarked:

> I typically only participate if I have a question pertaining to something that I don't know or I am trying to figure out. I think that there are a lot of White men typically participating. Honestly, in my major, I only know of about three African American women who are sophomores and have been in my classes. We don't participate unless we have a question in lecture.

Likewise, Carmen, a sociology major and former biology major, described her reasons for not answering questions in science classes as follows:

> Even now, especially as I got older, I got more and more introverted unless it was a social setting, and I cracked jokes. As far as class goes, now today, especially in college, I don't answer questions in class unless I am called on. And most of the times, I will choose to not answer questions in class. It really depends on the class that you are talking about. In the biology major, if you answer the question correctly, then they want to keep making you expand on the answer until they can stump you. As an example, in biology, if you answer something correctly, then the professors want to ask you a follow-up question and another follow-up question, so eventually you do not know the answer. And then it is, "Aha, you do not know the answer!" At least that is how I feel.

These stories indicate the levels of intimidation that African American women felt within their science classrooms at Town University, which contributed to their lack of course participation. Being one of a small number might also explain why some African American women are hesitant to participate in college science classrooms. To engage African American women in the college classroom, it is important to determine whether they have questions, so the instructor might ask the entire class if they have any final questions. If an African American woman raises her hand, then she should be called on. Otherwise, it is important for professors and instructors to be available during office hours to fully answer African American women's questions on science and math problems.

Modify Science and Math Teaching Styles in K–12 Settings

Teachers should engage African American girls in science and math classes by teaching to their learning styles in elementary and secondary settings using caring instruction and culturally responsive teaching. Elements of good teaching as defined by African American women as they reflected on their K–12 schooling included the following: (1) using multiple teaching methods (e.g., visual, auditory, kinesthetic), (2) reviewing course materials, (3) repeating course materials, (4) being relatable to students, (5) engaging students at every level, and (6) being interactive. For example, Briana, a biology major, asserted:

> A good science professor makes sure that the concepts are clear…and that each student understands those concepts. He has the ability to explain them in different ways for different students' learning styles. He can write exams that correctly reflect those concepts that he has taught in class, and students' exam scores correctly reflect their knowledge and his teaching ability through their scores.

Carmen, a former biology major who switched to sociology added to the discussion on good teaching. She stated:

> I think that there has to be a cohesive teaching strategy. I have seen so many teachers who have the class read the text, but they don't at all review the text in class. Probably, if you want us to read it and you are going to test us on it, you probably should go over it, reiterate it, because we may have not extrapolated what you feel like we need to know.

To Patricia, a business major, good teaching is person-centered. She explained:

> Good teaching is when you can relate to the students and teach the course materials in a way that you are teaching on everyone's level. For questions, you don't move too fast. You have different activities that will engage the class, and it is interesting. You make class fun.

Finally, engineering major Simone praised the teaching skills of her math instructor. She remarked:

> She was very interactive. She made sure that we understood the materials. She kept presenting them in an understandable way.

The comments presented in this section describe good teaching practices, such as making sure that students understand the course materials. Adapting classroom activities to accommodate students' varied learning styles is important as well. Good teaching also involves relating the content to the students' everyday lives, which is consistent with a culturally responsive teaching style (Gay, 2018). Professional development may be key so that K–12 teachers are informed about how to incorporate good teaching practices such as culturally responsive instruction in their classrooms.

Modify STEM Teaching Styles in College

Postsecondary institutions should make the science and math curricula and pedagogy accessible to nontraditional age STEM populations, such as women and minorities by using culturally responsive teaching (Gay, 2018) and culturally relevant pedagogies (Ashford et al., 2017;

Espinosa, 2011; Ireland et al., 2018). Faculty members should also consider adopting a caring approach inside and outside of the classroom that holds high expectations for students to succeed through learning in the classroom (Siddle Walker, 2001). By doing so, they can also serve as role models for African American women who want to pursue STEM fields.

The African American women in this study also identified elements of good teaching at the college level, including (1) use of technology to engage students (e.g., I-clickers) in question and response interactions, (2) more discussions and group work, (3) repetition of course materials, (4) interactive discussions, and (5) real-life applications of course materials. For instance, Raven, who switched her major to health science from biology, suggested that college instructors use a number of these approaches. She explained:

> The I-clickers are the quickest way to get an answer and response on how well your students are picking up on the materials. You can also do a bit more of discussions and group work in presenting the materials to the class. Oftentimes, students learn better when they have to restate the materials to someone else. Even the discussions can be a bit more interactive rather than the TA [teaching assistant] just repeating back what the professor already said a couple of days before.

Business major Patricia counselled against relying on lecturing from textbooks. Specifically, she said:

> I think we should stay away from textbook learning. It's not effective. There may be some things you have to learn from the textbook, but a lot of stuff that we learned is basically you have to learn it through experience. So, I like projects, because they give students

the opportunity to actually go out and do the work that they will be doing once they get into the profession. So, I think that's a good way to relate the course materials to the students and teach it in a way that it's not boring, but that we can understand.

Rachelle, a biology major, also identified several essential components of good teaching. She pointed out:

> Repetition, I think that is the best way for anyone to learn anything. So far, all of my teachers have a teaching strategy. I am not a fan of all of them. For example, I don't like my Neuro instructor, because he is unorganized. I think that organization and repetition are the best teaching strategies. Also, relating a concept back to everyday experiences always makes me remember something a little bit more.

Other African American women recommended using multiple methods of teaching (e.g., hands-on, visual learning, auditory) and making classes more interactive. These practices were specifically mentioned by Celeste, a sociology major and former pre-business major. She stated:

> People learn differently. You have people who are visual learners, people who are audio learners, people who are both. You have people who might need you to do a demonstration to learn how to do something, so it depends. It would be nice to have interactive things, like where students work in groups, and maybe students should have to do an in-front-of-the-class type of thing. I do not mind teachers calling on me, but make it a collective thing.

Crystal, a health science major who switched from biology, also emphasized employing different teaching styles to engage the entire range of student learners. She recommended,

> Try to put a little bit of everything in there. When I say that, I mean some of the class is hands-on, some of it is lecture, some is visual aids, because we have different people who learn in different ways. If you lecture all the time, the only people who are going to get something out of that are the people who can learn that way. And then if you do hands-on activities, the people who don't learn from doing will learn from hearing and seeing things. So, I would say, mix all of those into one.

Finally, math major Jennifer also supported making classes more interactive by having students "do more hands-on activities, instead of just talking at them. Ask the class to participate; ask students to come up to the board to provide answers." She further provided an example in the upper-level math classes:

> As you get in higher math classes, each problem may take an hour, so a good balance between providing practice for materials while at the same time not overly stressing out the students so that they still come into class and learn without crying every day.

African American women's suggestions for good teaching practices in college STEM majors are consistent with elements of good teaching practice in the literature on college STEM teaching (Harper, Weston, and Seymour, 2019). For example, Harper et al. (2019) found that good teaching practices for both students who persist in STEM majors and those who switch to non-STEM majors are as follows: (1) using interactive and inquiry-based teaching methods, (2) providing examples and real-life applications that make connections to the students' lives, (3) being organized, (4) being engaging, (5) being open and approachable, and (6) showing concern for the students. These practices are consistent with caring instruction (Siddle Walker, 2001) and culturally responsive teaching as well (Gay, 2018).

Expose African American Women to the Hidden Curriculum in STEM

To achieve success, African American women need to be exposed to the rules, norms, values, and behaviors that are expected of them within the STEM community because they may be the first individuals in their families to pursue a STEM degree. Consider, for example, Carmen's request to departments at Town University:

> Please provide students with the information that they need. Please provide them with the resources on campus; programs and organizations that they should be trying to get involved with. I know that in both of my majors, I get newsletters. Make the information that you want to provide actually interesting and accessible. Maybe you should schedule to meet with the students or mail them the newsletter so at least they actually see it. Promote things like internships that are coming up.

Math major Jennifer also suggested that departments provide more information to students. She stated:

> I feel like departments should provide more information on the classes besides just that little blurb description in the course guide. Provide the generic course guide information, but it might be a good idea to also provide a separate website for peers or even teachers themselves to write on that say things like "I give notes," or "You don't need to purchase this specific book." Or "This class is actually really simple if you

go to class," or "This class is actually pretty difficult, but the book is enough; you don't need to go to class." Those things sound maybe a little like giveaway things, but they really do factor into whether or not you take a class. You might sometimes be scared away from the classes, because you think "my other coursework is going to be too much. I can't also do that class." It might have ended up not being as rigorous as you thought, or the reverse will happen, a class will end up being more rigorous than you expected. I feel like information like that would be nice, just a little more detail available in a general sense, because otherwise, you have to know someone to find out more detailed information about classes.

In a similar fashion, African American women need to be aware of the culture of science at universities, especially at predominantly White institutions (PWIs). Several of the African American women provided tips for peers interested in STEM based on their lived experiences in science majors. Ashley, a biology major remarked:

I think that African American women in STEM majors need to be aware of the fact that there is probably a lot of doubt when people look at you; they don't just assume that you are going to be successful. They might assume that you are probably lost and wonder "why are you majoring in science?" and "why aren't you majoring in something else?" I just think that African American women in science have to know that if you work hard and sometimes you might have to work harder to get yourself out there for people to actually listen to what you are saying and for people to think that you know what you are talking about. But as long as you continue to work hard and just remain true to the fact that you are you and even though you might be outnumbered that does not mean that you have to stop or you have to feel that you are inadequate. You can be as successful as you want to be. There are definitely people who are always willing to help you no matter what you look like or where you come from.

Briana, also a biology major, added:

You are going to be one of the only Black women in your classes. Don't be intimidated by other people because you will be very intimidated. Just try to stay focused. Just study. Don't get discouraged. Even though it's hard not to get discouraged. Make sure you get help when you need it. Don't be afraid to ask questions. Don't be afraid to have fun, because you need to have fun; otherwise, you'll go crazy. Just like always, take time for yourself and just always know the limits of what you can and what you can't do.

In addition, biology major Shannon recommended,

Become organized, develop relationships with TA's [teaching assistant's] and professors. Develop relationships with your peers. I know I said I don't do the group study thing. It does help that I know people within my classes that I can call on if I do need some help for something or if we do need to work together. Just study; know your stuff. I know that there is a stigma on African Americans or any minority pursuing technical majors that they have this predisposition to not do so well. So, don't be discouraged. Don't be intimidated. Pursue it. Go hard for it and you will do fine! Keep the faith. If you are a Christian, then rely on that 90% of the time. That is a big support. If you don't have anyone in your life, that you can always pray. Or you can always go to church and uplift your spirits through that. Other than that, just don't let the social life overwhelm you. Stay focused!

Finally, sociology major Carmen shared the following thoughts for future African American women pursuing STEM majors in college. She stated:

> Before you come into this thing, you need to know your strengths and your weaknesses. Work on your weaknesses. You need to know what you can and cannot do. Do not think that you are invincible, because you aren't. Don't be so prideful that you can't ask for help. Seek help. You need to be aware that, while some exceptions are made for you, the odds are against you. They set aside seats in medical school or grad school for Black women. You are up against the odds. Fight, fight, fight! Fight to be recognized inside the classroom. Outside of the classroom, fight for the help that you need, money, or whatever you need.

These recommendations suggest that to be successful at PWIs, African American women need information regarding course requirements and the culture of science at their respective institutions.

Provide African American Women with Student Support Networks

Colleges should provide supportive spaces for academic and social engagement inside and outside of college classrooms. First, there should be a warm and welcoming space (Ashford et al., 2017; Ong et al., 2018), especially for African American women at PWIs, so that their needs (e.g., advising, tutoring, testing, studying with peers, building relationships with faculty members, dealing with discrimination based on race, sex, gender, or disability, preparing for medical school) can be addressed. This might include an African American women centered pre-medicine and STEM-focused student organization with a faculty advisor or administrator. This may also reduce some of the feelings of isolation because African American women will be working with a community of African American women within STEM fields.

Second, institutions should consider developing living–learning communities in dormitories for use by women of color studying STEM and pre-medicine during their freshmen and sophomore years. More specifically, living–learning communities should cater to both the academic and social needs of women of color in STEM majors (Johnson, 2011; Soldner et al., 2012; Ong et al., 2018). In the future, postsecondary institutions also might consider ways to promote family, school, and community partnerships to facilitate the retention, persistence, and graduation of African American women in STEM majors at PWIs.

Provide Warmer Departmental Climates for Learning in STEM

The final recommendation is that STEM departments should consider promoting warmer departmental climates among peers and faculty members to encourage collaboration with African American women and other underrepresented STEM and medical-school populations. Research has produced evidence on the benefits of warming up, that is, "the raising of students' initial aspirations after they enroll in a college" (Rosenbaum et al., 2009, p. 41). Warming up might include faculty members encouraging African American women to pursue STEM majors despite obstacles they encounter in four-year institutions.

However, instructors may need professional development to learn ways of showing that they care for their students. Thus, the first tip is to provide enthusiasm and caring instruction in class and in office hours. Consider the following recommendations to instructors provided by African American women in the study. Biology major Briana encouraged teachers to "just be open yourself. If you are just open and you help whoever wants help can come to office hours. Or if you just give off a very friendly aura about yourself, then students will come to you." She goes on to say "just make sure that you are eager to teach and really interested in whatever you are teaching about." This recommendation was echoed by Amber, a health science major, who contrasted the atmospheres of the biology and chemistry departments. She stated:

> Professors should be enthusiastic about what they want to do. I noticed in chemistry that the enthusiasm is there and people do better in chemistry. A lot of people change from biology to chemistry, maybe for that reason. They are trying to understand the materials, and the teachers are active and engaging. In biology, everyone is monotone and puts you to sleep. That should not be possible. And the way that they cram in so much material and talk in a soft voice. I do not see how they see that as learning.

In addition, math major Jennifer pointed out the importance of office hours in providing a warm and supportive environment. She said:

> Teachers should just really emphasize office hours more. I think some teachers are a little intimidating and maybe give off that aura of 'this is my office hour, but I don't want you to come.' I feel like the teachers

themselves should just reiterate, 'I will help you. Come to office hours, and I will help you. If you ask me questions, I will help you.' I've had some teachers who might seem really tough, and then after a few weeks, people in the classroom say, "oh, if you just go, he'll do the problem for you."

Shannon, another biology major, also highlighted office hours among her recommendations for providing a warmer departmental climate. She remarked:

Professors should just keep their doors open for questions. Hold office hours. Just make yourself available to students to answer questions. Also, have multiple resources. Not all students learn best from the online components. Not all students learn best in lecture. Not all students learn best by going to office hours. So, if instructors provide a setting that encompasses all of those things, then they can be meeting the needs of many students rather than just a select few.

This section provided recommendations that college instructors should be open and available to their students, especially African American women in STEM majors both inside the classroom and outside of the classroom during office hours. Instructors should also be enthusiastic and have a welcoming tone of voice that is not monotone, so that African American women are able to learn inside and outside of the classroom.

African American women in this study also desired to build personal relationships with their instructors, so it is important for college STEM departments to provide a space for that outside of the classroom. For example, Raven, a health science major and former biology major, described the difficulty of interacting with college instructors in the classroom setting. She said:

> Most times, science classes are really huge, so you really don't get that interaction. The only interaction that you would get is with the clicker. You can't really ask too many questions, because if you have 650 students trying to ask a question, then you will never get through the lesson. It is really hard to pinpoint how you could address student questions because there are so many students. It is hard to fit any in. Office hours tend to be more personable.

Patricia, a business major, provided some advice for instructors to be more caring for the needs of their students as learners. She stated:

> Instructors should just get to know their students on a more personal level and not just strictly as a teacher, but teach in a way that they could relate to the students and show that they're there, even outside of the classroom.

Chemistry major and former a health science major, Kayla also expressed an interest in greater interactions with her professors. She said:

> In addition to office hours, they should—I don't know if this is asking too much—make their presence more known in the study sessions. I know they have review sessions, but that is the entire class. I don't know if it is done by request for like a small group of people who are struggling. They should request that they have an extra study session that is guided by the professor. They should provide review sheets as well. I know that for my math class, there are some TA's who make up review sheets and give them to their students. I always

end up with the TA's who don't. So, I feel like those students have more of an advantage than those who don't have a review sheet. I feel like everyone should have review sheets.

Biology major Rachelle further suggested that teachers "could offer more advice and reach out to their students" to share information on topics such as courses to take or future career paths.

In summary, these narratives suggest that there is a need to provide warmer departmental environments for African American women in terms of peers and faculty members who should consider practicing caring instruction to effectively retain African American women in STEM majors at PWIs, like Town University. Being present in office hours is important to helping African American women learn the concepts in science and math classes. Career advice and advice for navigating the major might be advantageous to share in the study sessions or office hours so that these African American women are fully prepared for the rigors of STEM majors and even STEM careers.

In addition to these changes, some adjustments in the climates of science and math departments are needed to support African American women pursuing STEM and pre-medicine courses of study. The departmental changes might include attitudes about weed-out classes and who can succeed in STEM. For instance, Carmen, a former biology major who switched to sociology, requested that STEM departments "be more supportive just overall." Addressing the departments, she reasoned:

> You let this person into this program, because they earned their seat. There has to have been something in there that you thought that they could succeed. Even if they fall on their face, the overall concept had

to be there in the beginning so that they could have made it.

In addition, biology major Ashley described some specific ways in which departments could be more supportive. She stated:

> Well, chemistry can stop designing exams to make people fail. I don't believe in weed-out classes, so I hope that practice ends at some point. I definitely think that professors should let students know that there are places where they can get help from TA's, the professor, and tutoring. They should just let students know that they don't have to do it on their own.

Along the same lines, Crystal, a health science major and former biology major, also encouraged a more supportive departmental climate. She remarked:

> Let everybody know that they have the same ability to succeed in the class. One thing I do remember about my biology professor, on the first day, he said, "Everybody won't make it through this class. A lot of you will drop." That was on the first day, in the introduction of the class. It was just, like, "Well, thanks for encouraging me." I would prefer for a professor to say, "You all can succeed if you do what you need to do"

Other recommended departmental changes include providing resources to African American women pursuing STEM majors, such as study group support inside the classroom and tutoring and personal support outside of the classroom. For instance, regarding study groups, Kayla, a chemistry major and former health science major, suggested, "I would say definitely assign the groups inside the discussion instead of letting people choose their own, because those who are most quiet and reserved are going to end up with each other and that does not help at all."

In addition, health science major Danielle commended the personal support she experienced at a different university as a model. She explained:

> Professors should act like they want to be there. I feel like every student will want to be there. It's just like that enthusiastic spirit about everything. If I could take you to the class that I was in at Midtown University (pseudonym) They all had a willingness to help. And even in my application process, there were more than enough people emailing me so many times. They would say, "We want you here." "Do you need help with this? We'll help you out." And I wasn't even at the school yet. It's just like these people, they kind of love you already.

Health science major Amber recommended that tutoring be provided in STEM departments. She stated:

> Have tutoring available for all levels. The tutoring programs only go up to a certain level, but it does get harder. There have to be other ways to understand the class materials and get it in such a short amount of time. I think that more money needs to be put into those resources and programs instead of buildings.

Finally, Lara, a math major, provided the following suggestions:

> Offer more resources, especially to students. I know that they have other resources like math competitions and math organizations but usually you have to have a certain GPA [grade point average]. They should offer more resources without GPA

obligations. Something for students in math. Have workshops and study tips to help them within the major. They don't do enough of that. I get a lot of announcements through my biology and chemistry classes about extra stuff going on outside of class, even if it is a minute outside of the lecture. Or have people come in and talk about things going on campus that are looking for math majors or chemistry or biology majors. Or have people interested in these things go talk to people at schools.

These recommended changes indicate that to be retained within their majors, African American women need support from their departments, including attitudinal changes for faculty and staff members to advocate for their success. Resources, such as study groups and tutoring should be provided as well.

Finally, to change the departmental climate for African American women, institutions should also consider hiring departmental academic advisors and counselors who understand the personal, academic, social, and financial needs of African American women and can encourage them to persist in STEM majors despite barriers. Academic advisors can also facilitate the success of college students by encouraging students to strengthen their study skills (Young-Jones et al., 2013) and seek out campus resources, such as tutoring, student success centers, chemistry, physics, and math resource centers, peers, and instructors for academic success.

Students can also be referred to financial resources, including scholarships and student financial aid offices. Moreover, there may also be a need for professional development in advising to help advisors avoid cooling-out African American women in STEM majors and instead learn how to warm them up to fully engage in the rigors of STEM majors at PWIs and encourage them to pursue medical degrees and careers in STEM fields. Other resources include career centers and offices that help African American women cope with personal challenges, such as psychological services and student disability services. In conclusion, collective support networks and institutional support will increase the educational attainment of African American women in college-level STEM degree programs at PWIs while simultaneously preparing them to serve as leaders and change agents in industry in the United States and globally.

Conclusion

In conclusion, this study offers several recommendations for helping African American girls and women to succeed in K–16 schooling in STEM fields. First, cultivate African American girls' interests in science and math by exposing them to math and science projects in primary and secondary school settings. Second, engage African American girls in math and science by using culturally responsive teaching. Third, employ caring instruction using multiple teaching styles to meet the learning needs of African American girls and women in science and math classrooms in K–16 settings. Fourth, expose African American women to the hidden curriculum in STEM, including the academic rigor and competitiveness of courses, course failure and information overload, attendance at office hours, making use of campus resources, and departmental culture. Fifth, provide African American women with student support networks (e.g., student organizations, living–learning communities) that cater to their academic and social needs. Sixth, adapt the culture of

STEM departments to be warmer towards African American women in terms of attitudes among peers, faculty members, and staff regarding their abilities to succeed in STEM majors.

References

Ashford, S. N., Wilson, J.A., King, N. S., & Nyachae, T. M. (2017). STEM SISTA spaces creating counterspaces for Black girls and women. In T. S. Ransaw and R. Majors (Eds.), *Emerging issues and trends in education*. Lansing: Michigan State University Press.

Espinosa, L.L. (2011). Pipelines and pathways: Women of color in undergraduate STEM majors and the college experiences that contribute to persistence. *Harvard Education Review, 81*(2), 209-240. https://doi.org/10.17763/haer.81.2.92315ww157656k3u

Gay, G. (2018). *Culturally responsive teaching: Theory, research, and practice*. New York: Teachers College Press.

Harper, R.P., Weston, T.J., & Seymour, E. (2019). Students' perceptions of good STEM teaching. In E. Seymour & A. Hunter (Eds.), *Talking about leaving revisited: Persistence, relocation, and loss in undergraduate education* (pp. 245-276). Switzerland: Springer

Ireland, D. T., Freeman, K.E., Winston-Proctor, C.E., DeLaine, K. D., Lowe, S. M., & Woodson, K. M. (2018). (Un)hidden figures: A synthesis of research examining the intersectional experiences of Black women and girls in stem education. *Review of Research in Education, 42*, 226–254. https://doi.org/10.3102/0091732X18759072

Johnson, D. R. (2011). Examining sense of belonging and campus racial diversity experiences among women of color in STEM living-learning programs. *Journal of Women and Minorities in Science and Engineering, 17*(3), 209-223. doi: 10.1615/JWomenMinorScienEng.2011002843

Ong, M., Smith, J. M., & Ko, L. T. (2018). Counterspaces for women of color in STEM higher education: Marginal and central spaces for persistence and success. *Journal of Research in Science Teaching, 55*(2), 206-245. https://doi.org/10.1002/tea.21417

Rosenbaum, J. E., Deil-Amen, R., & Person, A. E. (2009). *After admission: From college access to college success.* New York: Russell Sage Foundation.

Siddle Walker, V. (2001). African American teaching in the south: 1940-1960. *American Educational Research Journal, 38*(4), 751-779. https://www.jstor.org/stable/3202502

Soldner, M., Rowan-Kenyon, H. Inkelas, K. K., & Garvey, J. (2012). Supporting students' intentions to persist in STEM disciplines: The role of living-learning programs among other social-cognitive factors. *Journal of Higher Education, 83*(3), 311-336. https://doi.org/10.1080/00221546.2012.11777246

Young-Jones, A. D., Burt, T.D., Dixon, S., & Hawthorne, M.J. (2013). Academic advising: Does it really impact student success? *Quality Assurance in Education, 22*(1), 7-19. https://doi.org/10.1108/09684881311293034

Chapter 9

Conclusion and A Call to Action for African American Girls and Women to Keep on Dreaming in STEM and Medicine

To conclude, I will return to the initial research question, "What are the lived experiences of African American women who remain in science, technology, engineering, and math (STEM) majors and those who switch to social science and health science majors?" The answer is that African American girls have a true love for science and math in elementary, middle, and high school. They not only engage in math and science through projects, but they teach those subjects to their peers as well. The use of culturally responsive teaching and caring instruction as teaching methods in the primary and secondary school classrooms helped them to learn and engage in math and science.

In college, African American women had some challenges in STEM majors at Town University, a predominantly white institution (PWI). This was primarily due to the hidden curriculum, which created racial and gender inequalities that limited the number of African American women who were successful in STEM majors. The African American women who learned to navigate the hidden curriculum in STEM fields became successful in obtaining bachelor's degrees at Town University. In addition, the collective support of peers, parents, professors, religious institutions, and student organizations through using social capital helped both African American women in STEM majors and those who transitioned to social science and health science majors to be successful in college.

At Town University, the cooling-out process within the STEM culture pushed out African American women who aspired to earn not only STEM degrees, but to become medical doctors as well. As a result of the current STEM culture at Town University, many African American women do not have equal opportunities to learn and pursue their dream STEM majors or even careers as medical doctors. Therefore, PWIs that have cooling-out processes in STEM majors need to engage in institutional changes within the STEM culture if they not only want to retain, but graduate more African American women with STEM bachelors' degrees.

This chapter concludes with a call to action for parents, K-12 teachers, college professors, college administrators, and mentors to help African American girls and women to keep on dreaming about pursuing STEM degrees and medical degrees! African American girls and women deserve to have equitable opportunities to learn in STEM majors, but pursue their dream undergraduate and graduate majors, medical degrees, and careers as well. This in turn can increase the pipeline of African American women scientists, technologists, engineers, mathematicians, and medical doctors from PWIs.

Epilogue:
Case Profiles and Where are They Now

This chapter provides context on each of the cases in the book. It offers a biographical sketch of their lives prior to attending Town University. This includes, the neighborhood contexts, school contexts, career aspirations, parental education and occupations. Finally, each case ends with an update on the education and occupation of each participant (if provided).

Case 1: Regina

Regina was a soft-spoken and reserved African American woman. During childhood, she recalled being interested in business at the tender age of eight after playing with cash registers. At this time, she had an infatuation with business suits, briefcases, and cell phones as well. Moreover, throughout the interviews, she continued to deliver safe answers, which was perhaps due to being an introvert. This point was illustrated through her narrative on the family context. For example, she recollected being the youngest child who grew up in a single-parent family comprised of her mom and three siblings. Family gatherings consisted of meal preparation at her mother's home for holidays like Christmas. Although she came from a lower-middle class family, her family traveled to the East Coast, West Coast, and the South for vacations.

Regina lived in Statesville from age five to a young adult. She lived in a family centered neighborhood in Statesville. She remembered Statesville being a diverse city. However, the U.S. Census Bureau (2000) reported that Statesville comprised of 65% Whites and 35% minorities in the 1990s. The average household size was four people; and about 60% of women and men were married as well. Additionally, 65% of people ages 16 or older held jobs. Furthermore, the median household income for families in Statesville was $65,000 between 2005 and 2009 (U.S. Census Bureau, 2009). Likewise, less than 10% of the families lived below the poverty line in her city. Furthermore, in Statesville, 90% of the people in her town held a high school diploma and 35% earned a bachelor's degree.

Consistent with Statesville's population, Regina's mother obtained a bachelor's degree. In fact, she graduated with a teaching degree in biology (a hard science field) from a minority serving institution. Her sister also completed college. At the time of the study, her brothers worked on degrees at technical colleges. Similar to her family, the majority of her neighborhood peers went to college with the exception of one peer who became a teenage mother. The majority of her public high school friends attended college as well. This was understandable, due to the fact that her predominantly minority, suburban high school offered advanced placement and honors curriculum (National Center for Education Statistics, 2010). Growing up, she participated in the National Junior Honor Society in middle school, National Honor Society in high school, and Business Professionals of America.

In 2008, Regina entered Town University as a freshman interested in a pre-business major. After the first year of college, she took off a semester from Town University due to financial concerns. Then she enrolled at a community college. At the time of the study, Regina was a junior majoring in sociology with a minor in African American studies. During the spring 2011, she took sociology classes for her major and African American studies classes for her minor. Her main obligations were school and working with women's organizations on campus. In the fall 2011, she was excited to report her upcoming graduation from Town University in spring 2012. In June 2020, Regina reported graduating from Town University with a Bachelor's degree in Sociology in 2012. Since undergrad, she pursued a Master of Education degree in Counseling with a concentration in College Student Development. She is currently completing Yoga Teacher training.

Case 2: Briana

Briana was an African American woman with a bubbly and outgoing personality. She aspired to become a doctor as a child. She always came to the interviews with a smile on her face in spite of her personal and academic challenges. To illustrate, she was enthusiastic when describing her science interests. She also remembered being interested in science around age 7 or 8 after attending a special space program at a research intense university for gifted and talented elementary school students. Similar to her openness in describing her science interests from day one to the last interview, she was very open and honest. For instance, she revealed that she was adopted by a working-class African American foster family in Central City Heights. She resided with them from birth to age 15. In fact, she paralleled her life story to the movie, *The Blind Side*. She recalled her neighbors being African American and Mexican in Central City Heights. She described Central City Heights as diverse.

Growing up with the African American family, she recollected family gatherings and travel that consisted of leaving town to other Midwestern states or the South for funerals. However, her life changed at age 16 when she transitioned from her African American foster family into a children's shelter. Then, she moved to Townville Heights to live in a foster home with lower-middle class White parents and their son. Similar to Briana's description of Central City, according to the U.S. Census Bureau (2000) Central City Heights was a diverse city with a population comprised of 60% minorities and 40% Whites. In addition, the family household size was three and a half people. The marriage population was about 40% of men and women as well. Furthermore, approximately 60% of people 16 or older were employed in the 1990s. Likewise, the median household for families in Central City Heights was less than $50,000 (U.S. Census Bureau, 2009). To add to that, less than 10% of families lived under the poverty level. In Central City Heights about 75% of the population ages 25 or older held a high school diploma while less than 15% earned a bachelor's degree as well.

Consistent with the population of Townville Heights both of her White foster parents (now both deceased) earned high school diplomas. Additionally, her White foster brother graduated from college. However, her foster sister from her Black family never attended college. This was congruent with the educational attainment of people from Central City Heights. Similarly, some of her neighborhood peers from Central City Heights and Townville Heights matriculated at community colleges. However, none of these friends attended four-year institutions. On the contrary to Briana's neighborhood peers, she went to a racially-mixed, suburban public high school with a gifted program and honors and advanced placement curriculum to prepare students for college (National Center for Education Statistics, 2010). As a teenager in high school, she participated in the National Honors Society, speech team, a volunteer organization called the Key Club; she played badminton, and was the volleyball manager.

In 2008, she entered Town University as a biochemistry major. As a junior in spring 2011, she majored in biology and minored in chemistry. She took biology, chemistry, theatre, and statistics classes as well. Outside of classes, she engaged in the organization Relay for Life; volunteered at a food bank, engaged in research in a lab, and worked 15 hours between two jobs. In fall 2011, she was excited about her upcoming spring 2012 graduation. In June 2020, she reported graduating from Town University in the spring 2012 with a Bachelor of Science degree in Biology. After graduating from college, she found it difficult to find a job, so she took some courses towards a second bachelor's degree in a clinical lab program. Since then, she took courses to become a physician's assistant. Currently, she is working on a clinical lab license. She began her career working in a paint factory. The next position that she worked in was in a science laboratory as a research lab technician. Her next role was as a research assistant/manager in a hospital system. After that role, she worked in a private pathology lab as a scientist working on biopsies for patients with lymphoma and leukemia. Currently, she works as a scientist in a Cancer Center. She continues to dream about becoming a doctor or a physician assistant one day.

Case 3: Shannon

Shannon was an African American woman who wanted to become a doctor. She was a soft-spoken, friendly, and assertive African American woman. She developed an interest in science in fourth grade due to her instructor and exposure to science labs. Equally as important, Shannon was a reserved person in the interviews. However, she shifted into a more open and honest person during the third interview on her college experiences. For example, she described the challenges studying for science and math classes, due to the fact that she never studied in high school. Similarly, she disclosed being the youngest of the three children in a lower-middle income two-parent household. She also recalled family gatherings centered on meeting with relatives during the holidays and summer in her home. As a child, she traveled with family members to the West, West Coast, and the South as well. Additionally, from her teenage years to early adulthood, she traveled with family members to the South. She also went on cruises in the United States and overseas with relatives as well.

She was born in Central City. She lived there up to 12 years-old. At that time in Central City, she described her neighborhood as being African American, peaceful, and quiet. She also associated with Black female friends who lived in middle-class, single parent homes, blended families, or two-parent households. Around age 12 her family moved to the suburbs of Redford Heights. She observed the neighborhood to be diverse with younger and older African American families. To complicate the neighborhood context, the U.S. Bureau (2000) reported that Central City had a majority White population in the 1990s. The average family household size was three people. In addition, 50% of men and women were married. Moreover, approximately 55% of the population ages 16 or older held jobs. The median income of families was less than $50,000. Less than 5% of families lived below the poverty level as well. Furthermore, 75% of people in Central City held a high school diploma and less than 10 percent obtained a bachelor's degree.

Consistent with the Redford Heights educated population (U.S. Census Bureau, 2009) both of Shannon's parents earned high school diplomas. Her mother also completed a two-year Associate's degree in Nursing (a health science field). In addition, her father attended college, but never finished due to military obligations. Occasionally, her older brothers took college classes. Similarly, the majority of neighborhood kids (approximately 15 to 20) enrolled in college. Likewise, based on the Redford Heights population, Shannon was destined to attend college. Prior to college, Shannon attended a low-income, predominantly minority, suburban high school (National Center for Education Statistics, 2010). Shannon's high school offered several advanced placement science courses and a handful of advanced placement courses in math, English and social studies.

Growing up, she was a part of the National Junior Honor Society in middle school. In high school, she worked 35 hours at Office Max. She was also on the gymnastics team, a student ambassador, Mathlete, a majorette; and she was a part of the Business Professionals of America. After high school in fall 2009, she became the first person in her family determined to complete a four-year degree in biology at Town University. At the end of spring 2011, she finished up sophomore year as a biology major. She took a chemistry, biology, psychology and a human development class. Outside of classes, she participated in an international organization that fundraised to build clinics for children. She also served as the caregiver of children when the teen mothers participated in support groups. More importantly, she sacrificed working a job to focus on her major during her freshman and sophomore years of college. However, in fall 2011, she continued engaging in the biology major, worked at a job, and participated in leadership roles in student organizations at Town University. In May 2020, Shannon reported successfully graduating in summer 2013 with a Bachelor of Science degree in Biology from Town University. However, she left the field of STEM and pursued a career in Human Resources. She may return to STEM as she is considering pursuing a Master of Business Administration in Information Systems.

Case 4: Carmen

Carmen hailed from Panama. She identified as an African American woman who wanted to become a doctor. She was a very talkative person. For instance, she was very enthusiastic when telling her story about her science interests. To illustrate, she described her leaf collection in grammar school. She also discussed her interests in science in sixth grade after being exposed to an oncologist inside a middle school classroom and her mother's diagnosis with cancer. She also grew up as the youngest child in a working-class, two-parent household with one biological brother and cousin who she considered to be a brother. Her family traveled across seas to visit family members, but they rarely took family vacations, due to their social class. It was not until high school when her parents traveled on cruises, because they moved from a working-class to lower-middle class social background after completing college degrees.

She remembered being the only Black family in a predominantly White neighborhood in South Town. However, she recalled that over the years the demographics of the population changed to being more violent with a diverse population. Consistent with her description South Town was 80% White and 20% minority (U.S. Census Bureau, 2000). Additionally, the average household size for families was three people. In South Town, 55% of males were married when compared to 45% of females as well. Moreover, approximately 70% of the population ages 16 and older were employed in South Town from 2005 to 2009 (U.S. Census Bureau, 2009). South Town also had a median family income of less than $60,000. Less than 10% of families lived below the poverty line as well. Furthermore, about 90% of the people in South Town held a high school diploma while less than 25% of people ages 25 or older earned bachelor's degrees. Similar to South Town's population Carmen's parents completed high school diplomas. In addition, her mother finished a Bachelor's degree in Nursing (a health science field) and her father completed a Master's degree in Civil and Mechanical Engineering (a hard science field). Unlike her parents, her brother did not attend college. Additionally, the majority of Carmen's neighborhood peers refused to matriculate in college. However, she remembered the big celebration of one African American female peer that graduated from college.

Equally as important was that Carmen encountered several challenges in schools, which ultimately led to her receiving home schooling up to 3rd grade. She later attended private schools. In grammar school, she was a part of the cheerleading team and dance team. Then, she went to a private high school. After that, she attended a racially-mixed, suburban high school that offered a handful of advanced placement courses (National Center for Education Statistics, 2010). She also ran track; she was also on the debate team and pom squad. In that school, students earned college credit from the neighboring community colleges or four-year institutions as well. So, after high school Carmen attended a community college. After acquiring better study habits, she matriculated at a private predominantly White college down South. Then, she transferred to a Historically Black College and University, Brownstone University due to the chilly racial climate at the other private White college.

In spring 2010, Carmen transferred to Town University as a junior. At that time, she was a biology major preparing for medical school. A year later in spring 2011, she was a senior majoring in sociology, because she was unable to fully complete the biology requirements at Town University. For the spring 2011 term, she elected classes including: an international health policy class, geography of health care, and a social theory class. She also worked a full-time job averaging 40 hours per week and volunteered at a hospital. In December 2011, she graduated from Town University with a Bachelor of Art's degree in Sociology.

Case 5: Ashley

Ashley was a biracial woman who identified as an African American woman. She aspired to be a singer, musician, a doctor, and a lawyer as a child. Ashley was a confident, yet reserved and soft-spoken biracial woman. She identified as an African American woman, mainly because of her Black Cuban father, but her mother was White. She also became interested in science in high school. During that time, she engaged in more laboratory experiments, because her high school had excessive resources unlike her previous schools. She was also very open during interview one when describing her family background and neighborhood contexts. For example, she grew up as the middle child of three children in a working-class, two-parent household. She recollected family travel consisting of visits to the Midwest to see family members. Outside of that, her family rarely traveled due to their social class standing.

In addition to the family context, Ashley's neighborhood context contributed to her schooling experiences. To demonstrate from newborn to age 6, she described her neighborhood as diverse with Asians, African Americans, Whites and young families on the West Side of Central City. Then from ages 7 to 8, she lived in a predominantly Black neighborhood on the West Side of Central City. Today, her parents live in a low-income neighborhood comprised of Blacks, Indians, and Pakistanis in Central City.

Furthermore, she discussed that there were gang members, gun shootings on a regular basis, and teenage mothers in the neighborhood.

The U.S. Census Bureau (2000) also reported that the West Side of Central City was diverse with 50% Whites and 50% minorities. In addition, the average household size for families was three and a half people. Less than 35% of men and less than 45% of women were married as well. More importantly, approximately 80% of the residents ages 16 or older were employed on the West Side of Central City (U.S. Census Bureau, 2009). The median family household was about $45,000. Less than 10% of families lived below the poverty line as well. Another significant statistic was that approximately 75% of people held high school diplomas and 20% acquired bachelors' degrees on the West Side of Central City.

Comparable to Central City's educational attainment, Ashley's parents obtained high school diplomas. Although her parents did not complete college degrees, all of her siblings attended college. Additionally, she had one family member who finished a degree in Agricultural Sciences (a hard science field) at a four-year institution. Similar to her neighborhood and family background, she recalled that one neighborhood peer attended and graduated from college. However, the educational choices of Ashley's parents provided her with tools to successfully transition into Town University as a biology major. Consider the fact that Ashley attended a private White, college preparatory high school that offered multiple advanced placement courses to prepare students for college (National Center for Education Statistics, 2010). In high school, she was part of the African American club; she also played on the basketball team as well.

In 2007, she entered Town University as a freshman interested in biology. In spring 2011, she was a senior majoring in biology. She enrolled in three dance classes and a biology class. She also engaged in laboratory research, participated in Black student organizations, worked on a senior thesis, exercised, and devoted 10 to 20 hours per week for the job in a dining hall. In the fall 2011, she entered a PhD program in Biology at a research intense university. She continued her studies in the PhD program in spring 2012. In May 2020, Ashley reported completing the Ph.D. in Biological Sciences. She is now working as a research project manager at a private research intense university.

Case 6: Jennifer

Jennifer was an outgoing and talkative African American woman. She became interested in math in fifth grade, in part due to her successes with the math problem of the week. Plus, her social and friendly personality might have contributed to her being open from interview one when describing her family background and neighborhood contexts. For instance, she discussed being the oldest child in a middle class two-parent family, which comprised of her mother, father, and brother. She also stated that she was a daddy's girl while her brother was closer to her mom. Moreover, she remembered family gatherings with her mother and brother, because her father was absent due to traveling for work. Her family traveled to the West, West Coast, East Coast, Midwest, and the South. Outside of the United States, she traveled to Mexico and the Bahamas with family members as well. Beyond the family context, Jennifer's neighborhood context might have accounted for her smoother transition into predominantly White school contexts. More specifically, she resided in a Jewish suburban neighborhood. She described her neighborhood in Southwest City being predominantly White. However, Southwest City was racially-mixed with 55% minority and 45%

White residents in the 1990s (U.S. Census Bureau, 2000). The average family size was three people. Forty percent of the married population consisted of men and women. Additionally, about 60% of the population ages 16 or older held jobs. This might have contributed to the median family income of less than $45,000. Furthermore, fewer than 25% of families lived below the poverty line. However, in Southwest City 70% of people held a high school diploma and less than 20% obtained a bachelor's degree.

Inconsistent with Southwest City's population, Jennifer's parents were highly educated with advanced degrees. Her mother has a Bachelor's degree in Marketing (a hard science field) while her father has a Bachelor's degree in Architectural Engineering (a hard science field) and a Master's in Business (a hard science field). So, consistent with the parental education, Jennifer (a college senior) and her brother (a college sophomore) attended college. Similarly, she recalled that all of her neighborhood peers attended college. Likewise, Jennifer's parents sent her to predominantly White schools in K-12 settings. As a consequence, she remembered being one of the few African American girls in her classes from elementary school throughout high school. However, the private high school that she attended offered multiple advanced placement courses to prepare students for college (National Center for Education Statistics, 2010).

Hence in 2007, she began her journey in college as a mathematics major at East Lake University (pseudonym), a predominantly White institution on the East Coast. Consequently, two years later in 2009, she transferred from East Lake University to Town University, due to homesickness, causing her to miss her family. In spring 2011, Jennifer was a rising senior majoring in mathematics. She enrolled in mathematics, economics, and a physical science class. She was also a very busy woman who participated in organizations from honor societies to volunteer organizations. In May 2011, she graduated from Town University as one of the few African American math majors. By November 2011, she was working as a financial advisor at a company. So, she remained working in a STEM community post-graduation. Currently, she is employed as a financial planner. She has also earned a Series 7, Series 66, Life and Health license, and a CFP®.

Case 7: Celeste

Celeste was a soft-spoken African American woman who once aspired to become a doctor. Her interest in a career as a doctor waned after dissecting a pig and squid in school. Similar to describing the science interests, she was open when discussing her family and neighborhood background. Celeste claimed that she came from a working-class, single-parent family household headed by her mother, due to her parents' divorce. She was the middle child of six as well. She also remembered family gatherings for holiday meals (e.g., Thanksgiving, Christmas, and New Year's) and/or funerals with her mother and five siblings (e.g., three sisters and two brothers). When her father was in the home, the family traveled to Disney World when she was around the tender age of 10 or 11 years-old. Her parents divorced after that trip. This resulted in fewer family trips due to financial reasons. However, school trips enabled Celeste to travel with peers to Midwestern states, the South, and the East Coast.

Celeste described the African American neighborhood filled with crime, teenage mothers, and poor or working-class single-parent mothers with boyfriends living in their homes. In contrast, her mother was considered to be doing better than other mothers, due to being employed and her marital status with a husband with a two-person income. However, her mother later divorced her father and then became a single-parent mother raising six children. Comparable to Celeste's description of her city, according to the U.S. Census Bureau (2000; 2009) the West Side of Central City has been less than 50% White and more than 50% minority from the 1990s throughout 2009. The average family household size was three and a half people. The majority of men and women were not married. Consider for example that less than 35% of men and 45% of women were unmarried. In addition, about 80% of the population ages 16 or older worked. However, the median household family income was approximately $45,000 in 2009 (U.S. Census Bureau, 2009). Less than 10% of families lived below the poverty line as well. Moreover, on the West Side of Central City approximately 75% of people ages 25 or older held a high school diploma while about 20% of people earned a bachelor's degree.

Similar to the West Side of Central City's population Celeste's mother and father earned high school diplomas. However, neither her mother nor father completed a college degree. On the contrary, her aunt earned a Doctorate in Nursing (a health science field). Despite this educational success in her family, she remembered that fewer than 20 people in her neighborhood attended college. Unlike her neighborhood peers, the educational choices that Celeste's mother made might have contributed to her preparation for college. Consider for example that instead of attending the neighborhood high school, Celeste's mother sent her to a private predominantly minority high school, which prepared 100% of students for college (National Center for Education Statistics, 2010). She was on the cheerleading squad, softball team, and Future Business Leaders of America (FBLA). Although this school offered multiple team sports, the school only offered a handful of advanced placement and honors courses.

Despite the limited number of college curriculum offered to students in high school, Celeste entered Town University with an interest in Business in 2006. Suddenly by the end of freshman year, Celeste realized that the demanding curriculum at Town University differed from her high school. She neglected studying in high school. So, she stopped-out for a year from Town University. During that year she went to massage therapy school. By 2008, Celeste reentered Town University as a sophomore eager to find a major for herself. In spring 2011, she was a rising senior majoring in sociology. She took a criminology class, a labor employment class, and an educational psychology class. Since she had scholarships, she was unemployed to focus on her heavy course load of 21 hours. She discussed that her daytime schedule consisted of going to class, doing homework, sleeping, and eating. In May 2011, she became the first person in her immediate family to graduate from college. In our follow-up meeting in the summer 2011, she admitted the desire to pursue a doctorate. By December 2011, she finished up the first semester of a Master's program in Sociology at a public research institution. In May 2020, Celeste disclosed that she completed the Master's degree in Sociology. She also had earned an Associate's degree in Nursing. She completed a postbaccalaureate program at a research intense university to strengthen her application to medical school as well. Currently, she is working as a nurse.

Case 8: Zaria

Zaria was case number 8 because of her willingness to engage in a phone conversation about the study. In the phone conversation, she gave oral consent to participate in the study. Zaria was omitted from the study, because I never obtained written consent from her to engage in the study. I engaged in a number of outreach techniques through calling and emailing her, but Zaria was unresponsive. Every time I attempted to fill the slot in case number 8 with another participant, the prospective candidate dropped from the study. So, I added an additional participant to the study. Zaria remained as case number 8 to acknowledge challenges with securing participants in qualitative research in the data collection phase. I also recognize Zaria's contributions to qualitative research even though she elected to withdraw from the study. Other qualitative researchers remembered participants who departed their studies as well (see Evans-Winters, 2003).

Case 9: Amber

Amber was an African American woman who wanted to become a chef and doctor as a child. She had a very outgoing and assertive personality. So, it was easy interacting with her during the nine interview sessions. In the second session, she discussed that her science interests began in sixth grade, due to her family members' health conditions. She continued to open up during latter interview sessions. For instance, in session three she discussed challenges transitioning into a racially-mixed high school and new town.

Equally as important was the family context. Amber grew up in a working class two-parent household with five younger siblings. She recollected family gatherings consisting of barbeques, large parties, holidays and family reunions. Her family traveled to the Midwest and the South as well. Besides the family context, Amber's neighborhood context was important to understanding her latter schooling experiences. She grew up in Southville Heights from birth to age 17. She remembered the racially-mixed population with Black and White residents. She also recalled a lot of kids in her neighborhood. However, she recollected that Townville Heights was predominantly White. Consistent with Amber's description of Southville Heights, about 20 % of people classified as White while approximately 80% identified as minority (U.S. Census Bureau, 2000). The household size was three people. About 70% of men and 55% of women were married as well. From 2005 to 2009, about 70 percent of people ages 16 and older were employed (U.S. Census Bureau, 2009). This might have accounted for the median family income of about $55,000 and fewer than 10% of the population living under the poverty line. Moreover, over 80% of the people in Southville Heights held high school diplomas and 25% obtained a bachelor's degree.

Comparable with Southville and Townville's (U.S. Census Bureau, 2009) population Amber's parents held high school diplomas. However, her parents earned associates' degrees at community colleges. By the end of the study, her mother graduated college with a Bachelor's degree in Nursing (a health science field) in summer 2011. In fact, Amber's mom was the first person in the family to attend and graduate from college. Plus, her father worked as a mechanical engineer (a hard science field). Additionally, the majority of her relatives on her father's side of the family completed degrees at Historically Black Colleges and Universities (HBCUs) in the South. A couple of her relatives hold nursing degrees and medical degrees as well.

Similar to her family's educational attainment, she recalled that the majority of peers from her first high school attended college. All of her friends from the second high school went to college as well. This might be attributed to the fact that the racially-mixed high school offered advanced placement and honors curriculum to prepare students for four-year institutions (National Center for Education Statistics, 2010). In high school, she was in the science club, band, the Spanish Club, and Mathletes. She played volleyball, ran track, and played badminton. She was also on the pom squad. More importantly, unlike some of her family members and school peers from Townville Heights, Amber was the first person in her family to attend college at a four-year research intense university.

In 2009, as a freshman, Amber was interested in majoring in biology to prepare her for medical school to fulfill the career aspiration as a medical doctor. She also wanted to create a dance studio to address the health needs of African American girls. So, for spring 2011, she took chemistry, a health science class, and a dance class. She also participated in a couple of organizations on campus and she worked part-time. She declared biology as her major at the end of spring 2011. By summer 2011, she was considering transferring into the health sciences as a major to prepare for medical school. However, during this transition phase, she worked and retook a biology class. In February 2012, Amber disclosed that she was a health science major.In May 2020, she reported that in December 2013, she was the first person in her family to graduate from a four-year research intense university, Town University with a Bachelor of Science degree in Health Science. Upon graduation, Amber had some difficulties obtaining a job in STEM, so she pursued a Master of Science degree in Biotechnology from a private university. She worked as a scientist in a research lab until Covid-19 happened in April 2020. She also discussed that she is continuing to pursue her dream of becoming a doctor.

Case 10: Raven

Raven was a dependable and outspoken Jamaican who identified as an African American woman who wanted to become a physician as a child. She came to meetings in sickness and in good health. Her dedication and openness began during day one. By interview session two, she described her interest in science stemming from working on a science project centered on the heart at age 8. By reflecting she acknowledged that project contributed to her passion and drive to pursue a career as a cardiologist. In a similar fashion, she opened up to describe her family background. She grew up in a working-class, single-parent household as the youngest of three children, two of whom are her biological brothers. By age five, she recalled that her brothers moved down South to live with their father. Suddenly, she became the only child living with her mom. As a consequence, for family gatherings, she remembered the family dinners with her mom to celebrate Thanksgiving and Christmas. Family travel involved her mother, neighbors, and her going down South as well.

Similar to the family context, Raven provided a vivid description of her neighborhood context. She remembered the violence of people, shootings, gangs and teenage mothers in her predominantly African American neighborhood in a Central City. Despite Raven's neighborhood demographics, Central City was considered a predominantly White city with a population of 80% Whites in the 1990s (U.S. Census Bureau, 2000). The average family household size was three people. About 50% of men and women were married at this time as well. Fifty-five percent of the population ages 16 and older were employed too (U.S. Census Bureau, 2009). However, the median family household income was less than $50,000. Less than 5% of the population lived below the poverty level as well. Furthermore, 75% of the population held a high school diploma and less than 20% earned a bachelor's degree.

Consistent with Central City's population, Raven's parents completed high school diplomas. Her oldest brother also attended college, but he dropped out due to family obligations. None of her family members completed degrees in science or math. Similar to her family's educational attainment, she remembered that an African American woman friend attended college from her neighborhood. She left college due to having a baby. So, Raven was the first person in her family and neighborhood to persist in college. Unlike the majority of her neighborhood peers, Raven attended a predominantly minority high school that offered advanced placement and honors classes in science, math, English and social studies (National Center for Education Statistics, 2010). These courses might have prepared Raven for her studies at Town University. In high school, she was a student ambassador; she was on the yearbook committee, dance team, and National Honor Society as well.

In fall 2008, Raven began as a biology major at Town University. Three years later in fall 2011, she was a rising senior in health science. She took courses, including speech communications, biochemistry, and health. In her free time, she volunteered at a hospital or for her student organization. She was also a mentor in student organizations, and held leadership positions in health organizations at Town University. Outside of student organizations, she worked at a hotel on the weekends. In spring 2012, she will be the first person in her immediate family to graduate from college. In spring 2012, she reported being accepted to a postbaccalaureate program to prepare scholars for medical school. In May 2020, Raven discussed her completion of the Bachelor of Science degree in Health Science. She also completed a MEDprep certificate program prior to entering medical school. She earned the Doctor of Medicine in 2018. Currently, she is working as a physician.

Case 11: Patricia

Patricia was a soft-spoken, yet assertive African American woman. Patricia was an African American woman who desired to be a teacher or a businesswoman as a kid. From the beginning of the interview sessions, she was an open and honest person. Consider for example, she spoke about always having an interest in math. In grammar school she remembered selling candy for a fourth grade class business project. Similarly, she was candid about her family background. She lived in a lower-middle class, extended family with her mother and grandparents as well. Although she had six siblings, she claimed to be an only child, due to the fact that she only lived with her mom. Her siblings lived with her father. As a result, she remembered family gatherings consisting of relatives congregating at her grandmother's house for holidays, like Labor Day, Memorial Day, Christmas, and Thanksgiving. Her family traveled to the Midwest and down South for family vacations as well.

In addition, for the first twelve years of her life, she grew up with her extended family on the North Side of Central City. She recalled it being a diverse city with Whites and African Americans. She also remembered being a part of an afterschool program with her African American neighborhood peers. In addition, from age twelve to the present she lived in the suburbs of Central City Hills with her mother. She described Central City Hills as diverse with Mexicans, Arabs, and African Americans, and Whites. However, the U.S. Census Bureau (2009) reported that about 70% of Central City Hills were White and 30% minority. The average household size for families was about three people. Approximately 55% of men and 45% of women were married at that time. About 70% of people ages 16 or older were employed as well. In addition, the median household family income was about $70,000. So, it was understandable that less than 15% of families lived under the poverty level. More importantly, the U.S. Census Bureau (2009) reported that 90% of the population held a high school diploma and less than 25% earned a bachelor's degree.

Similar to Central City Hills's population, Patricia's mother earned a high school diploma. However, her mother began college to pursue a degree in biology to become a dentist, but she dropped out of college. She also recalled that one aunt finished a degree in Cosmetology. None of her family members completed degrees in science or math. Comparable to her family background, few people in both neighborhoods that she lived in attended college. Unlike her neighborhood peers, Patricia enrolled at a predominantly minority college preparatory high school that offered multiple advanced placement and honors courses (National Center for Education Statistics, 2010). In high school, she was in the Entrepreneurship Club, student government, and on homecoming prom court. She in fact won homecoming queen/princess. She held leadership positions in student organizations, like the secretary of student government and served as vice president on the Executive Board of the National Honor Society. For the Yearbook Club, she was a general member and wrote for the sport's column.

In 2008, Patricia entered Town University as a freshman interested in business. Three years later in fall 2011, she was a rising senior at Town University. She took several business classes and an information technology class. She also worked two jobs on the campus of Town University. Additionally, she was involved in student organizations and mentored younger students. More importantly, in spring 2012 she will be the first person to graduate in her immediate family and neighborhood with a degree in Business.

Case 12: Simone

Simone was an African American woman who had a career aspiration to be the President of the United States of America. Simone was an assertive and talkative, African American woman with a bubbly personality. She came to every meeting cheerful and ready to talk despite her demanding schedule. For instance, she recalled that the science interest eventually developed in high school. She was open and honest in describing her high school experiences, which could be in part due to her vivid memories of those experiences. She remembered mixing chemicals and the reactions of those chemicals in the labs. Similarly, she was open in discussing her family background. She grew up in a middle class, two-parent family comprised of her mother, father, and brother. She also remembered that family gatherings comprised of food and fellowship with relatives for Thanksgiving, Christmas, the Fourth of July, Mother's Day and Father's Day. Family travel consisted of road trips to the South to visit family members, because her father despised air travel.

Once again, she provided a thorough description of her neighborhood, Town Heights. She described it as a suburb with young families. She also felt as though the suburb comprised of majority Blacks and fewer Whites. Consistent with Simone's description of Town Heights, the suburb was predominantly minority (U.S. Census Bureau, 2000). The mean family household size was three and a half people. It had a population with 55% married men and 45% married women as well. In addition, 65% of the population ages 16 or older were employed (U.S. Census Bureau, 2009). Moreover, the median household income for families was less than $65,000. However, less than 5% of families lived below the poverty level. In Town Heights about 90% of the population held high school diplomas and about 30% obtained a bachelor's degree as well.

Comparable with Town Heights' educational attainment, Simone's mother and father earned high school diplomas and college degrees. Her mom began her studies at Town University, but transferred to a different college. She also obtained an advanced degree and serves as a teacher. Her father went to a different PWI to complete a degree in computer science (a hard science field). On the contrary, her uncle completed a degree from a HBCU in the South. Similar to her family's educational attainment, she recalled that the majority of the neighborhood kids attended college.

In high school, Simone attended a predominantly minority high school. It offered advanced placement curriculum to students to prepare them for college (National Center for Education Statistics, 2010). So, it was understandable that Simone attended college given that she elected advanced placement courses in high school. She was a part of the math club as well. In 2007, Simone entered Town University as an engineering major. For fall 2011, she was a rising senior majoring in engineering. She enrolled in engineering and food science courses. She was also a member of an engineering organization. She sacrificed other student organizations to devote the remainder of her time to school in fall 2011. In May 2020, Simone reported that she graduated from Town University in summer 2012. Upon graduation, she worked as a project manager for a construction company. Later, she worked for a realty company. Currently, she is a Master of Business Administration student at a research intense university in the Midwest so that she can go into consulting, strategy, and marketing.

Case 13: Crystal

Crystal identified as an African American woman who wanted to become a doctor as a kid. Crystal was a talkative and friendly African American woman who came to every meeting with a smile on her face. She opened up in the first meeting discussing her participation in multiple organizations. Throughout the interviews, she continued to be honest about her educational experiences. For instance, she was always interested in science until her chemistry experiences in high school. She liked science because the curriculum centered on the body. In another interview session, she spoke about being the only child in a lower-middle class, single parent family until her mom married her stepfather. She also remembered family gatherings consisting of going to her grandmother's house for holiday dinners (e.g., Thanksgiving). She traveled with her family down South to visit family members and to the Midwest as well.

Similar to the family background, Crystal had a clear memory of her childhood surroundings. She described the Southern region of Central City as diverse. She recalled interacting with Whites, African Americans, and Mexicans in her neighborhood. Unlike her neighborhood in Central City, the statistics from the U.S. Census Bureau (2000) revealed that it was predominantly White in the 1990s. The mean household family size was three people. Fifty percent of men and women were married as well. Similarly, about 55% of the population ages 16 and older were employed (U.S. Census Bureau, 2009). However, the family household income was less than $50,000. In addition, about 10% of the population lived under the poverty level. Approximately 70 percent of the population ages 25 or older obtained a high school diploma. Less than 10 percent held a bachelor's degree as well.

Consistent with Central City's population, Crystal's mother finished high school and some college. Her father earned a high school diploma as well. She also revealed that an aunt obtained a bachelor's and master's degree from a minority serving institution in the 1980s. In addition, she spoke about a cousin who went to a PWI similar to Town University. However, none of her family members finished a degree in math or science. Similar to her family's educational history, only three neighborhood peers went to college.

Equally as important was that Crystal attended a predominantly minority and low-income, public high school. She also participated in a dance program. The high school offered a handful of advanced placement and honors classes in science, math, social studies, and English (National Center for Education Statistics, 2010). More importantly, the curriculum offered through the public high school gave her the chance to succeed in college at Town University. On the other hand, it might have contributed to her challenges in the science curriculum at Town University. Consider for example in 2008 when Crystal entered Town University, she was interested in pre-med and the biology major to become a doctor. However, the biology curriculum and class structure contributed to her changing to the health science major.

In fall 2011, she was a rising senior at Town University taking health science classes, an internship, and a horticulture class. She also worked at a job and mentored youth through her involvement in student organizations at Town University. In spring 2012, she will be the first person in her immediate family to graduate with a bachelor's degree. In June 2020, Crystal discussed that she completed the Bachelor of Science degree in Health Science from Town University. She also pursued her Master of Art's degree in

Health Communications at a private institution. Although she did not pursue the medical degree, she still wanted to work with patients. Her current field in Human Resources allows for her to advocate for them.

Case 14: Danielle

Danielle was an African American woman student athlete who wanted to become a chef and an actress on Broadway. Danielle was an outgoing and religious African American woman. Beginning in the first interview, she opened up to share her story. Consider for example, her interest in science began at an early age. She also remembered enjoying science over subjects like English and social studies. In addition, she was very candid about her family background. She was born in Central City and grew up as the youngest child in a middle-class, two-parent family with three kids. She also recalled family gatherings centered on spending time with relatives on holidays, such as Thanksgiving, Christmas, Fourth of July, and Labor Day. Her family traveled to the South and East Coast for business meetings or vacations as well.

Besides Danielle's family background, her neighborhood context might explain her latter schooling experiences. For instance, Danielle lived in Central City from birth to age 11 in a middle-class African American neighborhood. Then her family moved to Statesville, a suburb. Her parents still live there in 2011. In early adolescence, she remembered there being more Whites in Statesville than Central City. From adolescence to early adulthood, she recalled Statesville being a racially-mixed suburban town. She also recollected there being an African American family in her neighborhood similar to her family composition with the wife, husband, and their older children.

Consistent with Danielle's description of Statesville, the U.S. Census Bureau (2009) reported that about 65% of the population was White and 35% minority. The household size of families were four people. Approximately 60% of women and men were married as well. Additionally, 65% of the population ages 16 and older were employed between 2005 and 2009. Statesville also had a median family income of $65,000. Fewer than 10% of families lived under the poverty level. About 90% of the city's population ages 25 or older earned a high school diploma and 35% obtained a bachelor's degree as well.

Comparable with Statesville's educational attainment, both of Danielle's parents finished high school and college. Her parents hold degrees in Computer Science (a hard science major). In addition, her sister went to a small four-year institution. Recently, her brother enrolled in a community college after serving in the National Guard. Her uncle completed a degree in Kinesiology (a health science major) and an aunt finished a degree in Nursing (a health science major) as well. Similar to the educational background of her family, it was no coincidence that Danielle enrolled at a predominantly White, suburban high school. Her school offered multiple advanced placement and honors classes to prepare students for college (National Center for Education Statistics, 2010). As a teenager, she played on the women's basketball team and ran track. She participated in the Fellowship for Christian Athletes and a youth group as well.

In 2009, she entered Town University as a freshman majoring in biology. By fall 2011, as a junior, Danielle transferred into the health science major at Town University. She took courses in subjects including biology and health science. Outside of class, she worked a couple of hours at a job, participated in the track team at Town University, and attended church on a weekly basis. In June 2020, she reported that she earned the Bachelor of Science degree in Health Science in May 2013, a Doctor of Physical Therapy (DPT) in 2017, and a credential called the Certified Orthopedic Manipulative Therapist (COMT) in 2019. She now practices as a physical therapist in an outpatient setting.

Case 15: Kayla

Kayla was a reserved yet friendly African American woman who aspired to become a teacher as a kid. Kayla always had a smile on her face in spite of the challenges that she faced due to her tight schedule and heavy course load. She was open and honest during the interviews when discussing science interests, along with family and neighborhood backgrounds. For instance, she developed an interest in science at an early age. She loved math in elementary school. She remembered doing fractions in grammar school as well. In addition, she reported that the social class background that she grew up in was a working-class foster family with five siblings (three brothers and two sisters). She remembered that the family gathered for reunions in the South. There, they had fish fry's and barbeques. Leaving the state for family reunions down South were considered family travel and vacations for Kayla.

Besides her family background, she recollected her childhood neighborhood being predominantly Black on the West Side of Central City. However, in the 1990s, the West Side of Central City was 80% White and 20% minority (U.S. Census Bureau, 2000). The average family household size was four people. Less than 60% of men and women were married. In addition, over 70% of the population ages 16 or older were employed (U.S. Census Bureau, 2009). The West Side of Central City also had a median family income of $65,000. So, it was understandable that less than

10 percent of families lived below the poverty level. Fewer than 75% of the population held a high school diploma and about 25% of the residents ages 25 and older earned a bachelor's degree as well.

Likewise, Kayla's foster parents graduated from high school. They also went to college, but never finished their degrees. Despite these educational challenges, Kayla's foster mother's granddaughter went to an HBCU down South. However, none of her family members completed a degree in math or science. Similar to Kayla's family's educational attainment, few of Kayla's neighborhood peers went to college. Of the ones who attended college, the majority of them enrolled at community colleges or less selective four-year institutions. Unlike many of the neighborhood peers, Kayla attended a private, predominantly White college preparatory high school with multiple advanced placement and honors classes to prepare her for college (National Center for Education Statistics, 2010). In high school, she was in the African American Club. In fact, she became the president of the African American Club as well.

In 2009, Kayla entered Town University as a freshman interested in majoring in health science. By fall 2011, Kayla was a health science major transitioning into chemistry as her major at Town University. She enrolled in science classes, including biology and chemistry and a Calculus II class. Outside of class, Kayla was a busy lady who worked as a residential advisor in a dorm at Town University. She also participated in organizations centered in chemistry and health. In May 2020, Kayla reported completing the Bachelor of Science degree in Chemistry from Town University. She also received her Doctor of Pharmacy in 2018. She plans to pursue a Master of Business Administration in the future. She began her career as a scribe in an emergency room, because she planned to become a nurse practitioner.

Her next position was in a hospital pharmacy. Currently, Kayla is a managed care pharmacist.

Case 16: Rachelle

Rachelle was a kind, talkative, and assertive African American woman. As a child, she wanted to become either a police officer, anesthesiologist, or go into neurology. She came to the second interview to discuss her experiences in K-12 schools and her mother's death, which occurred while Rachelle was in high school. As an example of her school experiences, she became interested in science in high school, and realized that the other subjects (such as history) did not interest her. She also described her family background. She identifies as middle-class, due to her parents' income, and grew up as an only child. She disclosed that her father remarried after her mother died, and that was when she gained a stepbrother. She remembers family gatherings with her uncles and aunts as well, which consisted of traveling to the West Coast and South for her birthday. She traveled to museums and malls in Central City.

Rachelle provided a vivid description of her former neighborhoods. She was born in Central City, and recalled the population as predominantly African American, with few whites. She remembered living in a quiet neighborhood within a gated community, although she reported that the quality of the neighborhood later deteriorated. Now, her parents live in a suburb called Northern Heights. She described the people there as less friendly, and the residents as predominantly white. However, she stated that her own neighborhood was largely African American. The U.S. Census Bureau (2009) reported that it was 75% white and 25% minority between 2005 and 2009, consistent with Rachelle's description of Northern Heights. According to the U.S. Census Bureau (2009), the average family household size was three people, with the marriage population about 60% for men and women. Sixty-five percent of people aged 16 years and older were employed; so, it was understandable that the median household family income was approximately $90,000, and that less than 5% of families lived in poverty. Additionally, approximately 95% of the population held a high school diploma, and about 40% of the population had earned a bachelor's degree.

Typical of Northern Heights' educational statistics, Rachelle's stepmother finished college with a bachelor's degree. Her father completed an associate degree, and later obtained a construction certification. Many of her neighborhood peers became teenage parents, and did not further their education beyond a high school diploma. However, she reported that about 90% of her graduating high school class went to college. She attended a predominantly minority high school which offered advanced placement and STEM courses (National Center for Education Statistics, 2010). In high school, she participated in the Book Club, Speech Team, Science Club, Key Club, and Spanish Club. She was also a cheerleader and a member of the National Honor Society.

In 2008, Rachelle entered Town University with an interest in biology. A senior in fall 2011, she then took chemistry and biology classes, while also teaching a service-learning course. She worked two jobs outside of schoolwork: in one position, she managed a team of students; in the other, she developed programs for freshman. She also participated in various student organizations, including an African American health club and a mentoring organization. Rachelle planned to graduate from Town University with a bachelor's degree in Biology, and reported in June 2020 that she did graduate in the summer of 2012 with a Bachelor of Science degree in Biology. In the future, she wants to pursue a Master of Business Administration degree. After graduation in 2012, she worked as a microbiologist, and later as a project manager in the chemistry industry. Currently, she works as a scientist, specifically as a project manager in the area of Research and Design.

Case 17: Lara

Lara was a soft-spoken yet assertive and helpful African American woman who aspired to be a doctor growing up as a kid. She was always interested in math. At the tender age of eight, she began working on math problems with her brother who was two years older than her. In addition, she opened up to describe her experiences in predominantly Black schools in Central City. For instance, she spoke about the multiple African American female math instructors in primary school throughout high school. The exposure to these role models might have accounted for her interests and success in math in K-16 schools. Moreover, she spoke about living in a lower-middle class, single-parent family with her mother and brother. She also remembered family gatherings consisting of food, fellowship, and fun during birthdays and holidays including: the Fourth of July, Christmas, and Thanksgiving. Her family traveled to Jamaica, Australia, and Mexico as well. Some of the traveling was in part due to her brother and her karate tournaments.

Lara was born and raised in Central City. She described Central City as predominantly White with a few Blacks and Latinos. She recalled her neighborhood peers coming from low-income families. She also remembered her peers participating in street life activities, which contributed to fewer of them graduating from even high school. Some of her neighborhood female peers became mothers as well. Consistent with Lara's description of Central City, the U.S. Census Bureau (2000) reported that 80% of the population was White and 20% were minorities in the 1990s. The average family household size was three people. The marriage population consisted of 50% of men and women as well. In addition, 65% of the population ages 16 and older held jobs (U.S. Census Bureau, 2009). The median family household income was less than $50,000. More importantly, in Central City, approximately 80% of the population held a high school diploma and less than 35% earned a bachelor's degree as well.

Comparable with Central City's population, Lara's mother obtained a high school diploma. Her mother also earned a bachelor's and a master's degree in Social Work (a social science field) from Town University. Her aunt obtained a degree in physical therapy as well. However, none of her family members earned degrees in math. She also recalled that few of her neighborhood peers attended college. The peers who enrolled in college took classes at community colleges. Unlike her peers, Lara's low-income, predominantly minority high school offered a few advanced placement and honors classes in English, math, social studies, and science (National Center for Education Statistics, 2010). In high school, she was a part of the Math Club and National Honor Society; she played tennis as well.

In 2009, Lara entered Town University as a math major. As a junior in fall 2011, she remained a math major. She enrolled in classes in subjects, such as biology, ethnic studies, health, and math. She also worked in the dorm to advocate for minorities and participated in student organizations. The student organizations, included an African American health organization and a mentoring organization. She was also a part of a Diversity Leadership Program. By the end of the study, she was considering switching her major to chemistry. However, in January 2012, she reported continuing to pursue a degree in math at Town University.

In May 2020, Lara disclosed the completion of the Bachelor of Science degree in Mathematics from Town University in the spring 2013. She earned a Doctor of Medicine in 2017. She has not pursued residency yet. However, she began a career in finance. She is still in the field of finance today. She would like to go into the medicine policy side of health care in the future.

Conclusion: Keep On Dreaming!

From these brief biographical sketches, we learned that all of the African American women graduated from Town University, which is a tremendous accomplishment in itself. They kept on dreaming of graduation despite obstacles while pursuing science, technology, engineering, and math (STEM) degrees. We also learned that in terms of advanced education that there were five doctors (2 Doctor of Medicine, 1 Doctor of Pharmacy, 1 Doctor of Physical Therapy, 1 Doctor of Philosophy). In terms of careers, there were two African American mathematicians, three research scientists, and three doctors practicing in their respective fields. Two of the African American women went into the field of human resources; 1 went into the field of higher education; and 1 went into nursing. To conclude, African American women should continue to dream about advancing in their education and careers despite obstacles.

References

Evans-Winters, V. (2003). *Teaching Black girls: Resiliency in urban classrooms.* New York: Peter Lang.
National Center for Education Statistics. (2010). Common core data for secondary schools [Data set]. http://nces.ed.gov/ccd/
U.S. Census Bureau. (2008). Industry business patterns [Data set]. http://censtats.census.gov/
U.S. Census Bureau. (2009). 2005-2009 American community survey [Data set]. http://factfinder.census.gov/
U.S. Census Bureau. (2000). Census 2000 Demographic profiles [Data set]. http://factfinder.census.gov/
U.S. Department of Education & National Center for Education Statistics. (2010). 2010 College Enrollment Statistics 2010 [Data set]. http://nces.ed.gov/

Appendix

Table 3: Overview of the Participants

Table 3

Overview of the participants

Name	Age	First College Major	Final College Major	College Standing
Regina	20	Pre-Business	Sociology	Junior
Briana	20	Biochemistry	Biology	Junior
Shannon	19	Biology	Biology	Sophomore
Carmen	22	Biology	Sociology	Senior
Ashley	21	Biology	Biology	Senior
Jennifer	21	Mathematics	Mathematics	Senior
Celeste	22	Pre-Business	Sociology	Senior
Amber	20	Biology	Health Science	Junior
Raven	20	Biology	Health Science	Junior
Patricia	21	Business	Business	Senior
Simone	22	Engineering	Engineering	Senior
Crystal	21	Biology	Health Science	Senior
Danielle	19	Biology	Health Science	Junior
Kayla	20	Health Science	Chemistry	Junior
Rachelle	21	Biology	Biology	Senior
Lara	21	Mathematics	Mathematics	Junior

McPherson, E. (2012). *Undergraduate African American women's narratives on persistence in science majors* [Doctoral dissertation, University of Illinois at Urbana-Champaign]. Illinois Digital Environment for Access to Learning and Scholarship. P.90

Table 3

Overview of the Participants Continued

Name	Mother's Education	Mother's Occupation	Father's Education	Father's Occupation
Regina	Master's Degree	Teacher	H.S. Diploma	Truck Driver
Briana	H.S. Diploma	Housewife	H.S. Diploma	Landscaper
Shannon	Associate's Degree	Registered Nurse	Some College	U.S. Steel Security & Fire Services
Carmen	Bachelor's	Nurse	Master's	Firefighter
Ashley	Some College	USPS Worker (mail carrier)	H.S. Diploma	City/Water Department
Jennifer	Master's Degree	Homemaker	Master's	Entrepreneur
Celeste	Some College	USPS Worker (mail carrier)	Some College	Veteran's Affairs Hospital Worker
Amber	Bachelor's Degree	Unemployed	Some College	Unemployed
Raven	H.S. Diploma	Supervisor (Hotel)	H.S. Diploma	Retired Carpenter
Patricia	Some College	Retired	Some High School	Deceased
Simone	Master's Degree	Teacher	Bachelor's Degree	Systems Analysis
Crystal	Some College	Retired	H.S. Diploma	Self-Employed Painter
Danielle	Bachelor's Degree	Computers	Bachelor's Degree	Computers

Table 3

Overview of the Participants Continued

Name	Mother's Education	Mother's Occupation	Father's Education	Father's Occupation
Kayla	Some College	Unemployed	Some College	Unemployed
Rachelle	Bachelor's Degree	Pharmaceutical Sales	Associate's Degree	Auto Worker
Lara	Master's Degree	Director of Mental Health Care Center	Unknown	Deceased

McPherson, E. (2012). *Undergraduate African American women's narratives on persistence in science majors* [Doctoral dissertation, University of Illinois at Urbana-Champaign]. Illinois Digital Environment for Access to Learning and Scholarship. P.91

Biography

Dr. Ezella McPherson earned a Master of Education and Doctor of Philosophy in Educational Policy Studies at the University of Illinois at Urbana-Champaign. She received her bachelor's degree from the University of Michigan-Ann Arbor. As an African American woman, she is familiar with the STEM culture through first-hand experiences, observations of friends, and former students in STEM majors and pre-medicine. She has also published manuscripts on African American women's giftedness in STEM ("Oh you are smart: Young, gifted African American women in STEM majors"), spiritual capital ("Having our say in higher education: African American women's stories of 'doing science' through using spiritual capital"), resilience ("African American women's resilience in hard science majors"), informal learning ("Informal learning in SME fields for African American undergraduate females"), and commitment to STEM majors ("To commit or leave from STEM majors at a PWI: An exploration of African American women's experiences"). She recently published the number 1 best-selling book titled, *Real Outreach: A Practical Guide to Retaining and Graduating College Students*.

Her co-authored manuscripts include the mentoring of African American women in STEM at Historically Black College and Universities ("Mentoring our own: African American women in engineering" with co-author Dr. Virginia Tickles) and examining the experiences of minority women in STEM ("Minority women in STEM: A valuable resource in the global economy" with co-author Dr. Diane Fuselier-Thompson").

She has presented on college student persistence, retention, and graduation of African American women, minority students, and science, technology, engineering, and math (STEM) students at conferences, such as the American Educational Research Association, American Educational Studies Association, Association for the Study of Higher Education, STEM Education Conference University of British Columbia, International Congress of Qualitative Inquiry, Oakland University Student Success Conference, American Sociological Association, and the Equity within the Classroom Conference. She can be reached via email for speaking engagements or workshops at emcpher2@gmail.com.